D0496193

Mayday! Mayday!

HEROIC AIR-SEA RESCUES
IN IRISH WATERS

Lorna Siggins

Cork City Library
WITHDRAWN
FROM STOCK

GILL & MACMILLAN

Gill & Macmillan Ltd
Hume Avenue, Park West, Dublin 12
with associated companies throughout the world
www.gillmacmillan.ie
© Lorna Siggins 2004
0 7171 3529 2

Index compiled by Cover To Cover
Print origination by O'K Graphic Design, Dublin
Printed by ColourBooks Ltd, Dublin

This book is typeset in 10.5pt/12pt Berkeley Book.

*The paper used in this book comes from the wood pulp of managed forests. For every
tree felled, at least one tree is planted, thereby renewing natural resources.*

All rights reserved.
No part of this publication may be copied, reproduced or transmitted in any
form or by any means, without permission of the publishers.

A CIP catalogue record for this book is available from the British Library.

1 3 5 4 2

348038

The illustrations on pages 32–5 by Ian Commin are reproduced from
Rescue: The True-Life Drama of Royal Air Force Search and Rescue by Paul Beaver
and Paul Berriff, published by Patrick Stephens Ltd, 1990.

CITY
LIBRARY
CORK

CONTENTS

Acknowledgments   vii

 1 Rescue from Roaninish   1
 2 The Invincibles   6
 3 Cliff-face challenges   23
 4 From Whiddy to west of Fastnet   45
 5 Beyond the limit   58
 6 The west coast campaign   67
 7 Close calls   82
 8 Heroes unsung   95
 9 *Dunboy's* near-ditching   102
10 Lost at sea — the crew of the *Carrickatine*   112
11 Sea caves and Clare cliffs   122
12 Calm patch, building . . . Oh Jesus Christ!   132
13 Tramore, Co. Waterford, 2 July 1999   138
14 Missing man formation   152
15 Flying blankets and the big man from Armagh   167
16 Skerd rocks survivor   175
17 Blacksod to Benbecula — crossing borders   188
18 The *Celestial Dawn*   201
19 Hours of boredom, moments of terror   211
20 Go mairidís beo   218

Notes   223
Glossary   227
Index   228

ACKNOWLEDGMENTS

Many have helped to write this book. I am grateful to *The Irish Times* for its support and wish to thank the following for their efforts and inspiration.

Fergal Tobin of Gill & Macmillan, who commissioned it, and Jonathan Williams, literary agent; Commandant Shane Bonner of the Air Corps and his colleagues who put in countless hours of research; Captain Liam Kirwan, director of the Irish Coast Guard; Captain Geoff Livingstone, his deputy at the Irish Coast Guard, and Eamon Torpay, Search and Rescue Operations manager; Tom McLoughlin, press officer at the Department of Communications, Marine and Natural Resources; Mark Clark, press officer with the British Maritime and Coastguard Agency; Esther Murnane of *The Irish Times* library; and Madeleine O'Rourke, journalist, archivist and pilot!

Individual pilots and crew gave generously of their time to research and explain particular incidents: Captain David Courtney, formerly of the Irish Coast Guard and Air Corps; Captain Derek Nequest of CHC Helicopters at the Irish Coast Guard, Shannon; Captain John McDermott, formerly of the Irish Coast Guard and Air Corps; Flight Sergeants Daithi Ó Cearbhalláin, Mick Treacy and Alan Gallagher of the Irish Coast Guard; Flight Sergeant Ben Heron; and current and former Air Corps officers and crew — Brigadier General Barney McMahon; Lieutenant General J. P. Kelly; Captain Chris Carey; Commandant Fergus O'Connor; Commandant Tom O'Connor; Captain Hugh O'Donnell; Lieutenant Colonel Harvey O'Keeffe; Commandant Paddy O'Shea; Lieutenant Anne Brogan; Lieutenant Lee Brennan; Captain Paul Hayes; Captain Andy Whelan; Sergeant Dick Murray; Airman Dermot Goldsberry; Flight Sergeant Christy Mahady; Commandant Sean Murphy; Commandant Dave Sparrow; Captain Dave Swan; and Commandant Jurgen Whyte.

I am indebted to RAF pilots, Squadron Leader Alan 'Rocky' Boulden; Flight Lieutenant Al Potter; Flight Lieutenant Ian Saunders; Flight Lieutenant Steve Johnson; and to Chris Ruhle of the British Embassy in Dublin. Many Air Corps, Irish Coast Guard, RAF, Royal Navy and British Coastguard air crews have performed vital and courageous rescue missions in these waters in co-operation with the RNLI, coast and cliff rescue units, mountain and coastal community rescue, and only a fraction of these are represented here.

Dr Marion Broderick GP, Aran Islands, Joan McGinley, former chair of the West Coast Search and Rescue Action Committee, and Joey Murrin, former chief executive of the Killybegs Fishermen's Organisation, went to great lengths to dig out information, while Arthur Reynolds and Bill Crampton trawled the archives of *The Irish Skipper*; maritime historian Dr John de Courcy Ireland kept meticulous records; Captain Brian Sheridan, Galway harbour master, and press colleagues Dick Hogan, former *Irish Times* Southern correspondent, Mairtín Ó Catháin, the *Connacht Tribune*, and Dr Muiris Houston, medical correspondent of *The Irish Times* gave of their expertise.

I am also grateful to Commodore John Kavanagh, Lieutenant Commander Gerry O'Flynn and Lieutenant Commander Gerry O'Donoghue of the Naval Service; Claire Brennan of the RNLI; Superintendent Tony McNamara of Ballyglass lifeboat in County Mayo; Lieutenant Commander John Leech of the Naval Service and Irish Water Safety; Jason Whooley of the Irish South and West Fishermen's Organisation; Lorcan Ó Cinnéide of the Irish Fish Producers' Organisation; Frank Shovlin; Donal O'Donnell, former skipper of the *Seán Pól*; mountaineers Dermot Somers, Maeve MacPherson, Breeda Murphy, Joss Lynam and Emily Hackett; Mairéad ní Fhlatharta; Alan and Chris Rintoul; Michael Fewer, author and hill-walker; and Ted Creedon of *The Kerryman*.

For use of photographic material, I am indebted to Peter Thursfield, Picture Editor of *The Irish Times*; Alan Betson, *The Irish Times*, Joe O'Shaughnessy, *The Connacht Tribune*; Liam Burke of Press 22; Patrick J. Cummins; Nutan; Michael Fewer; Frank Shovlin; the Air Corps photographic archive; Irish Coast Guard photographic archive; the British Ministry of Defence and RAF photographic section.

For their meticulous attention to detail, my sincere thanks to Gill & Macmillan managing editor D Rennison Kunz and photographic editor Aoileann O'Donnell; and to copy-editor and proof-reader Jim McArdle.

Cian Siggins, Daphne Siggins, Fidelma Mullane, Sybil Curley and Philly Eves provided the vital moral and practical support.

Finally, I must make special mention of Maria, Lily and Davina O'Flaherty, Tony and Mary Baker, Vincent and Anna Byrne and Monica Mooney, the relatives of the four Air Corps crew who lost their lives in the call of duty at Tramore, Co. Waterford, on 2 July 1999.

Lorna Siggins
February 2004

1

Rescue from Roaninish

When Nora Shovlin heard the roar, she thought it was the bus from Portnoo. Mick Meehan is very early this morning, she said to herself, as she rose and started cleaning the fireplace at her home in Narin in County Donegal. She was just putting out the ashes when she heard the noise again, and turned to see something she would never forget.

It was large, it was loud, it came from above, and it landed on the beach just below her house. A man in uniform emerged, head down, and deposited a green box on Narin strand. He waved to her, boarded, and the craft took off and headed towards Roaninish, a small island between Glen Head and Arranmore in County Donegal. Her five-year-old son John ran down to have a look; the box appeared to be empty and seemed to be serving as a marker.

It was 2 March 1956 and Mrs Nora Shovlin, then mother of six children, had just seen her first helicopter. Several hours later she would be cooking breakfast for its crew and for the ten men who were to be involved in one of the first recorded air-sea rescues on the Irish coastline.

Out on the rock, they sang, they drank what little they had and they cracked jokes. Picture a group of shipwrecked seamen — frozen, forlorn and huddled together on an outcrop in a storm in Donegal Bay. Their chances of survival seemed limited as the tide rose; their ship, the *Greenhaven*, had run up on the small island many hours before when its engine failed in bad weather the night before.

The *Greenhaven* was a 250 ton coaster owned by a company in Newcastle-upon-Tyne. Sean McKeon was a 34-year-old pilot employed to guide ships into Ballina Harbour in County Mayo over the bar in Killala Bay and into the River Moy estuary. On 1 March he had brought the *Greenhaven* in with a load of fertiliser and was asked by the captain to take it out again, although his colleague Mick Killeen was on duty.

The ship left that night with two pilots on board. Once it reached open sea both pilots were then due to have been picked up by a boat from Enniscrone, Co. Sligo, while the *Greenhaven* continued its passage to the next port of call, Belfast. However, conditions were too rough to make the transfer. A third pilot on board another smaller coaster, the *Galtee*, which was following in the *Greenhaven's* wake, was also caught by the weather. There was some consultation between the captains of the two ships and it was decided to take all three pilots to Belfast for safety.

The *Greenhaven* never reached the north-east, however. It was off Rathlin O'Beirne Island in Donegal when the chief engineer on board, Joe Champion, found himself frantically trying to repair the engine's cooling system. The *Galtee* tried to take the vessel in tow, and got a line aboard, but several attempts failed when two ropes and a hawser snapped in the high winds. The master of the *Greenhaven*, Captain Balmain, marvelled at the skill of his colleague from Listowel, Co. Kerry, Captain Paddy Histon, on the smaller *Galtee*. Unfortunately his seamanship was to no avail.

The engineer managed to restart the *Greenhaven's* engines, but they were sluggish and could only make two knots. Lacking both ballast and power, it was tossed about mercilessly in a rising swell and driven back gradually towards the coastline. The captain issued an SOS which was picked up by a British Navy frigate, the *Wizard*, and relayed to Donegal's Arranmore lifeboat. There was no navigational beacon on Roaninish Island as the vessel was thrown up on it, and the crew felt the sickening pounding and watched in horror as white water broke over the ship.

Sean McKeon remembers that all ten on board had gathered in the wheelhouse. 'We were like pigs in a lorry on a mountain road and all the glass from the wheelhouse windows was shattering in on top of us.' Over the next couple of hours the crew could hear exchanges between the Naval frigate and the lifeboat, but the captain was unable to acknowledge them as the receiver had packed up. 'We hauled out the morse signal lamp and started signalling the Navy ship,' McKeon recalls.

All the while, with every groan of the hull, they feared the vessel would slide off the rock at any minute and take them with it. 'Somehow we found our feet, and as the tide went out the ship broke in the middle and one mast fell to starboard and one to port,' McKeon says. 'We decided we had no future on the vessel and we had to get out.' The ship's mate and two seamen went down the rope ladder with the flash lamp at about 11.30 p.m. to check out the rocks. The highest rock, about half a mile from the ship, was a good 20 feet above sea level.

They returned to tell the rest of the crew and the two pilots. 'All the boys came down the rope ladder, some with biscuits, some with whiskey and gin in the bags,' says McKeon. 'It was heartbreaking because we knew there were cases and cases of drink on board!'

With them also were several flash lamps which they used to signal the *Wizard* and the Arranmore lifeboat when they saw it approaching Roaninish at about 2 a.m. The lifeboat planned to shoot out a line and take the men off by breeches buoy, one at a time. However, according to a subsequent report in *The Donegal Democrat*, at about 2.45 a.m. the Naval frigate's signalling light read: 'Ship destroyed. Seas too high to put lifeboat in.'[1]

By this time a plane was circling overhead and dropping flares to boost the men's morale. Most of them were soaked from the waist down and they tried singing songs like 'Sixteen Tons' to keep their spirits up. At around 4 a.m. the *Wizard* signalled, 'Food, clothing coming at daylight. Plane will try drop.' Balmain, who had been torpedoed twice during the Second World War, knew that the only hope was a helicopter. 'But look where we are, on the bloody west coast of Ireland,' McKeon remembers the mate remarking. Then someone thought of the British military base at Eglinton in Derry.

'Send helicopter' came the signal from the men on the rocks. The frigate relayed the request through Malin Head coast radio station to Eglinton and was able to respond: 'Helicopters here at eight. Don't worry.' There should have been whoops and cheers, but some of the crew had 'gone very quiet', McKeon remembers, spending any energy they had trying to keep warm. The gale was still blowing and they were being hammered by wind, showers of sleet and spray from the sea as the tide came in. 'We were crawling around like sheep and cattle on a mountain, but to tell you the truth we were very worried about the lifeboat,' McKeon says. 'It had stood by valiantly throughout the night to give us comfort, but the seas were really bad.'

The new day had dawned when another British military plane arrived and dropped parcels with bread, biscuits and blankets over the island. Some fell into the sea or were blown by the wind, but the men managed to salvage about half of them. They were still opening them up when someone spotted a black speck in the sky. The unmistakable roar overhead came at about 8 a.m. Within minutes a Royal Navy helicopter winchman was being lowered down on to the rocks, swinging on a fragile cable, while the men looked up at him with a mixture of fear and relief.

'He said he could only take two passengers at a time, so the first two were sent up and taken in to shore, and a second helicopter came out then and took another two,' McKeon says. 'That continued until we were all off.' He was one of the last to leave, with pilot Mick Killeen. The captain and pilot waited for the final relay. Only then did the Arranmore lifeboat leave its station to return to its home port, and the *Wizard* continued on its passage.

Nora Shovlin's house became the nucleus for all activity when the seamen were brought ashore. 'People had come down from everywhere to watch, and Mrs Shovlin and her husband Peter couldn't have been more welcoming,' Sean McKeon says. 'Mrs Shovlin cooked us breakfast and gathered clothes from the neighbours.' Her children stayed at home from school that day, as did many of their pals. The Shovlins' eldest son, Frank, who has written about the incident,[2] and who was at boarding school in St Eunan's in Letterkenny, read about these events several days later in the newspapers.

The helicopter crews were also made welcome in the Shovlin household as they had to wait for a tanker to come with fuel. The pilots were anxious to be airborne because they had no night navigation. The day's drama wasn't quite over, however, as they prepared to depart at about 10 a.m. 'With half of Downstrands looking on, the first helicopter took off and headed for Eglinton,' Frank Shovlin says. 'But when the second helicopter took off, it developed a fault and crashed into the field south of the strand. The crew sustained no major injury, but the helicopter was wrecked and had to be transported back to base on a 60 foot lorry which arrived next day and caused nearly as much excitement.'[3]

The Donegal Democrat quoted several eyewitnesses, including Garda M. J. O'Maille, originally from Galway city. 'She sounded quite normal as she rose. Next her engine spluttered out and I saw her drop. The engine came on again and she rose slightly and then it cut again. She started to fall . . .'

The newspaper also quoted Mrs Shovlin: 'We were standing at the front door here and were rooted to the spot as we watched it fall. We thought it was going to fall on the house, but it crashed just ten yards away from the doorstep. We thought she was going to burst into flames and shouted, but there was nothing more than a loud crash as she broke.'

The helicopter's wheels struck a hillock and it collided with a bank, severing the tail and shearing off its wheels on the nose. However, the three crew on board stepped out, shaken but uninjured, and were assisted by the local gardaí under Superintendent James McDonagh. The pilot was quoted as saying that he feared the gale would blow him on to one of the four houses nestled in closely together at Narin. 'We were quite powerless when the engine cut at that stage. Our altitude was 100 feet and our speed was very slow. If we had had 400 feet and a much faster speed, we would have been able to control it and glide in.'[4]

The gardaí had to provide the helicopter with overnight security. Frank Shovlin remembers that Garda Crowley spent the night sheltering in his father's haystack which stood close to the site of the crash. The receiver of wrecks travelled to Portnoo to interview Captain Balmain of the *Greenhaven*, and the crew found their various ways home.

The ship itself, lodged up on Roaninish, attracted much interest, but there was an unhappy sequel to the dramatic rescue, Frank Shovlin records. During the summer of 1956 three visitors were drowned between Roaninish and Portnoo. The two adults and a child who lost their lives were among a party which had set out on a sunny morning to view the shipwreck. *En route* home to Portnoo that evening, the weather changed and one of the boats foundered in choppy seas. George Warren, a solicitor from Enniskillen, Co. Fermanagh, Desmond McVitty, a businessman from Dublin, and Christopher Chambers, a seven-year-old boy from Belfast, lost their lives. The remaining eight in the party were rescued by an angling boat skippered by J. J. McBride, the principal at Clogher National School.

The *Greenhaven* continued to dominate Roaninish for over 30 years until it finally slipped into the Atlantic and beneath the waves.

2

THE INVINCIBLES

Shortly before Christmas in 1963 a French fishing vessel found itself in difficulties in heavy weather off the Connemara coastline. Its name was the *Emerance* and there were 16 crew on board when a distress message was sent out. Details were very scant but the vessel was believed to be drifting without engine power close to rocks and the crew had been forced to abandon ship and take to inflatable liferafts. Two trawlers working close by, the *Melchior* and the *Balthazar*, relayed messages to say they were on their way to assist.

Some 160 miles away on the east coast, an Alouette III helicopter took off from Baldonnel aerodrome in west Dublin. The pilot was Commandant Barney McMahon, a tall, young and enthusiastic Air Corps pilot from Doonbeg, Co. Clare, and this was his very first mission as detachment commander of Ireland's first air-sea rescue service.

Christmas trees were already up in thousands of houses below them as the helicopter flew west to Galway in deteriorating weather conditions. On board with McMahon were Lieutenant Fergus O'Connor and Sergeant Peter Sheeran. They had done some rough calculations and estimated they should be able to make it straight out to the Aran Islands without refuelling.

This was the theory at least, but the practice might prove to be a bit different. For it was less than a month since McMahon and his colleague Lieutenant John (J. P.) Kelly had delivered two new Alouette III helicopters, A195 and A196, from Marseilles in the south of France to Air Corps headquarters at Baldonnel. They hadn't established a regular training schedule in the Irish Sea, never mind fly over to the Atlantic seaboard.

'When I think of it now, we really were rookies, and I suppose I was a bit bloody mad to be taking that mission on,' McMahon, now Brigadier General and retired from the Air Corps, says. 'We had very limited communications and an atrocious radio which took the heads off us because we had no helmets, and we had 1916 lifejackets, no liferaft in case of an emergency, and the most basic of gear. We knew the Aran Island lifeboat had also been called out and so our plan was to try and make contact with it. But it took us ages to find it as it was heaving up and down in the swell — as were several fishing vessels near by.'

Back at base, arrangements were made to have extra fuel delivered to Renmore Army barracks in Galway. An Air Corps Dove aircraft was also dispatched to Galway to act as an airborne (top cover) radio relay station between the helicopter and Baldonnel. The Alouette battled its way west over Galway city and out over the bay towards the Aran Islands. Scanning the turbulent seas the crew could see nothing on the heaving grey and white surface below and could barely talk to each other with the noise of the engine and rotors. The pilot watched the fuel gauge constantly. There was still no sign of the *Emerance* or its crew in liferafts when they were forced to turn for land.

The aircraft wasn't going to make Galway city, and it was cruising in over north Connemara when McMahon spotted a likely landing place. Just outside the town of Clifden, he put the Alouette down in a handball alley. He needed aviation fuel (Jet A1), and fast, but he certainly wasn't going to get this in Clifden and there wasn't going to be enough time to wait for a delivery by truck, which could take almost six hours from Baldonnel. A few minutes later the proprietor of the Stella garage was surprised to see a pilot in military dress in front of him asking for 120 gallons of petrol!

The proprietor thought Christmas had come two days early — he hadn't sold that much petrol in a whole year. There was a problem, however; the pilot would need a certain amount of oil in the mix. The garage phoned the parish priest to see if he had any paraffin.

'Eventually we got sorted. We realised we had no filter, so I asked the garage owner if I could borrow a pair of ladies' nylons,' McMahon remembers. 'I think they were probably provided by his good wife, and they did the job.' Once the crew had enough fuel on board, they climbed into the aircraft, started up the engine and were airborne again in minutes. The staff of the Stella garage watched in some wonder as the machine rose up over the handball alley and flew out on a south-westerly course towards Inis Mór. They stood there, eyes fixed on it until it was just a small grey dot on the horizon.

'During that second search, we got a message to say the *Emerance* crew had been located,' McMahon says. One of the French vessels also searching the area that afternoon found the crew and picked them up. The vessel sent out a radio message which was picked up by a coast radio station in Brest in France and relayed it to the helicopter. It confirmed that all 16 crew were safe and sent a message of thanks to the Air Corps and the lifeboat for their efforts.

'We had been searching for four hours and 45 minutes in total when we turned back for Renmore again,' McMahon remembers. However, Clifden's handball alley became a frequent unofficial landing pad thereafter, and subsequently a fuel dump for Irish Helicopters. 'We got to know the nearest neighbours Ciaran and Lavinia Joyce quite well.' Ciaran was to become an agent for Irish Helicopters and his son Pat became a helicopter engineer and held an autogyro pilot licence. Pat Joyce remembers McMahon's landing clearly as it was the first time he and most of his contemporaries had ever seen a helicopter. 'We thought Santa Claus had arrived a couple of days early!' he recalls.

McMahon was glad he had Lieutenant (subsequently Commandant) O'Connor on board for that first mission. When he and J. P. Kelly had travelled to the Aerospatiale plant in Marignane, southern France, for their conversion training in early November 1963, they had watched with some amazement — and not a little horror — at the French technique for sea rescue. 'They would drop their winchman into the sea and he would swim to the cable which was then lowered down by the pilot. It meant that no winch operator was necessary as the pilot could see everything,' McMahon says. 'You are talking flat calm Mediterranean seas, of course. We knew there was no way we'd survive with that in the much more turbulent Atlantic.'

McMahon immediately contacted Colonel Billy Keane back at base and suggested that he send two officers to train with the Royal Air Force (RAF) in Britain. By the time the two Alouettes had arrived into Baldonnel on a dark and wet evening of 26 November 1963 after a historic two-day delivery run from Marseilles via Lyons, Paris, London and Cardiff, O'Connor and Corporal Jim Fahy had completed a four-week helicopter search and rescue crewman's course at RAF stations in Ternhill and Valley, and had some idea of how winching crews should work with pilots in hostile conditions in these waters. However, they hadn't had a chance to pass these skills on when the *Emerance* emergency arose. The first formal course for Air Corps crewmen/technicians — for the ground crew doubled up as winch volunteers in those early days — started on 31 December that year.

That delivery run of the new helicopters to Ireland almost put O'Connor to the test. He was having lunch on 23 November 1963 when he was summoned to an 'urgent' meeting at Air Corps headquarters. 'I was told that Barney McMahon and J. P. Kelly had just landed A195 and A196 in Cardiff Airport and had refuelled and received a Met briefing for the final leg to Baldonnel. Because of the strength of forecast headwinds, Barney felt that their fuel tanks might not be sufficient to fly direct to Baldonnel and that an *en route* fuel stop somewhere between Rosslare and Wicklow might be necessary,' O'Connor remembers. 'To cover for this, it was decided to dispatch the Air Corps fuel truck or "bouser" to the east coast with Sergeant Cummins and myself on board. Throughout its long service life the bouser had been confined to short refuelling trips at Baldonnel and was not really up to the journey planned. By the time it had chugged its way to Wicklow town, darkness was descending. I checked with the local Garda station and to my great disappointment was informed by the Garda sergeant on duty that about three hours previously he had seen two helicopters fly overhead towards Baldonnel. I telephoned Baldonnel and established that both helicopters had safely landed.'

O'Connor was more than a mite frustrated at missing the historic moment — and the ensuing celebrations. 'We turned for home and despite Ned's careful nursing and vocal encouragement, the bouser began to struggle. It took almost two hours to make the 20 miles to Delgany, where it finally wheezed to a stop — conveniently in front of the Delgany Inn. We telephoned Baldonnel for assistance and transport eventually arrived. I still expected to make the tail-end of the arrival party but, alas, only empty champagne bottles greeted me. The following day "words" were exchanged with my three colleagues, Barney McMahon, J. P. Kelly and Chris Carey.'

'AIR-SEA RESCUE HELICOPTERS go to aid of French trawler' read the headline in *The Irish Times* report of Christmas Eve 1963, recording the details of the 'newly formed' air-sea rescue service's first mission. 'The helicopter landed at Renmore barracks, Galway, and will fly back to Baldonnel today', the report concluded. 'It can fly on ordinary motor car petrol, but it has not the same range with this type of fuel as it would have with the proper grade.'

In fact, the helicopter had to be given a special engine check on its return to Baldonnel. And, as Captain Chris Carey — one of the first four Air Corps search and rescue pilots — recalls, there had to be 'a bit of a rethink' after the *Emerance* mission. 'The Alouette was far better than the RAF's counterpart at the time, but it was still a single-engine aircraft. I suppose in a sense we were all writing the operations manual. It probably helped that we used to whirl miraculous medals up into that engine from time to time!'

Hardly a month had passed when the young search and rescue crew were back out on the west coast again. Both McMahon and J. P. Kelly were on duty when the call came through that a pregnant woman was about to give birth on Inishturk Island in County Mayo. Would the helicopter be able to fly her to hospital in Castlebar?

McMahon, Kelly and Peter Sheeran weren't quite sure what was ahead of them as they flew north-west from Baldonnel to Castlebar, where they would refuel before flying on out to the island. Effectively this was their first air-ambulance mission and they were still training up for air-sea search and rescue. None of them had any suitable medical training and they had no doctor with them. If the woman gave birth during flight, would they be able to cope? But then, by this time they were the invincibles. 'We would have gone anywhere, done anything,' O'Connor remembers, 'and that was simply because we didn't have the experience to say no.'

The Alouette landed near the pier. There was no sign of anyone. They had been told that the woman, a Mrs Faherty, would be there waiting for them. They climbed out of the aircraft and walked around. The next moment McMahon remembers spotting a woman walking down towards the pier, carrying two enormous suitcases. She was very heavily pregnant. McMahon couldn't believe his eyes. Her husband, smoking a pipe, was walking empty-handed several steps behind her!

The couple climbed on board and the suitcases were heaved in beside them; if Mrs Faherty was in the advanced stages of labour, the pilot didn't remember her displaying much distress. 'We had to ask them to wear lifejackets, but we couldn't get it on Mrs Faherty due to her condition.' The flight to Castlebar Hospital took about 45 minutes, and there was at least one press camera present and several medical staff when the Alouette landed. As Mrs Faherty alighted, she turned to the pilot and handed him a £1 note. 'It took me a minute or two to realise this was a tip and I couldn't accept it — certainly not when there were cameras there,' McMahon says. 'She wouldn't take it back. She was insistent, and so I had to put it back into her handbag myself. Unfortunately that's what the camera caught — me with my hand in her purse! Goodness! Did I curse afterwards.' It was one of those moments that was to become part of search and rescue folklore back at the helicopter base.

THE RAF TRAINING WHICH O'Connor and Fahy had undertaken involved a team of winch operator and winchman, who carried a rescue strop first devised by the Royal Navy. Apart from looking after the winchman, the winch operator also had to guide the pilot while over a casualty or vessel by using a 'patter' or constant detailed dialogue. O'Connor and Fahy were all set to adapt this expertise to the smaller Alouette III helicopter and pass it on to the first group of non-commissioned officers (NCOs) selected for the new helicopter — Flight Sergeant Peter Sheeran, Sergeant Peter Smith and Corporals Mick Fitzgerald, Sean Oakes, Bill O'Connell and Liam Sheridan.

Funds were tight, however, and so the strop design was adapted at Baldonnel using old tyres which were tested out on various 'victims' suspended from a gantry in the Air Corps technical wing hangar. The first overwater exercise took place on 14 January 1964 with the Rosslare lifeboat in County Wexford. O'Connor's logbook records that Chris Carey was the pilot and he was winch operator. 'It took place within Rosslare Harbour, and that should never have been agreed to,' he says. 'The lifeboat crew had to duck to avoid the helicopter descending to a dangerously low height while trying to maintain a hover overhead, and it clearly indicated our complete inexperience as a search and rescue crew.'

On 25 February 1964 a launch was hired by the Department of Defence from the Royal St George yacht club in Dun Laoghaire, south Dublin, and four days later the first deck winching course took place. By early March the Air Corps was confident enough to demonstrate its new winching prowess at the Dublin Boat Show in the Royal Dublin Society grounds at Ballsbridge. Eventually immersion suits arrived in August and 'wet' winching training began; a Neil Robertson stretcher acquired from the Royal National Lifeboat Institution in October 1964 added a new dimension to the exercise programme.

However, by November, a year after the Alouettes' arrival, the Department of Defence was already belly-aching about the financing of this search and rescue service. It refused to agree a tender to supply a suitable training launch for the helicopter crews in the

Dublin area, and so any deck training that took place from late November to the summer of 1965 had to do so at the Naval base in Haulbowline, Cork Harbour, where the commanding officer had supplied the use of his launch, the *Colleen*. The cost of one helicopter flight to and from Cork for training would have paid for several months' boat rental on the east coast. As for the aircrews themselves, their enthusiasm was certainly not reflected in their pay packets: winching, which was regarded as a hobby for the technicians who volunteered to do it, earned a bonus of around £3 a year!

The new helicopters were seven months old when the first actual rescue at sea was carried out. 'Helicopter to the Rescue. Night drama off Howth: Man and boy plucked from the sea' read the page one report in the *Evening Herald* of 8 August 1964. 'In a dramatic five-minute rescue operation, an Army helicopter plucked a Dublin car salesman and his 12-year-old nephew from the sea this morning after a 16-hour ordeal adrift in a rowing boat off Howth last night', the *Herald* staff reporter wrote. 'The rescued man, Mr Frank Kellett, a cousin of the noted show-jumper, Miss Iris Kellett, gave a graphic description of the sea drama, today, at his residence, 163 Whitehall Road, Terenure.'

Barney McMahon was alerted at about 7 a.m. A small boat with two crew on board was in trouble around the Kish lightship in the Irish Sea, he was told. The Howth lifeboat under coxswain Joseph McLoughlin had been out searching since shortly after the alert was raised by Mr Kellett's wife at 12.30 a.m.

Mr Kellett and his nephew, Michael Maybin from Ballymena, had set off from Howth after tea to go sea angling in a rowing boat with an outboard engine. They got caught in strong currents and Kellett started the outboard engine which 'worked for a while' before packing up. When the boat began to drift out to sea Kellett began to row desperately to try and keep sight of land. He rowed for over five hours while his exhausted nephew slept in the bottom of the boat. On several occasions he thought he saw lights — the spotlights of the lifeboat — and he tried lighting papers to act as flares. At one point he was within 150 yards of several fishing vessels, but they didn't see him in the darkness.

The *Evening Herald* reported that Mrs Kellett waited out at the harbour in a Garda squad car as the lifeboat searched. 'I couldn't stay at home,' she said. 'I waited out there all night expecting him to be brought in dead at any moment.'

The helicopter had been out searching for about three-quarters of an hour when it located the drifting boat about a mile south-west of the Kish lightship. On board with McMahon were Airman Dermot Goldsberry and Corporal Bill O'Connell. Airman Donal O'Leary was on Howth Head to communicate with the helicopter via a radio set. Winch operator O'Connell spotted the boat first and Goldsberry went down the cable and put the rescue strop on the young boy. When he was winched up the airman returned for the boy's uncle.

'Both of them were frozen and the sea was just a small bit choppy,' McMahon remembers. The actual rescue took about five minutes and the helicopter then flew to Baldonnel. Lightmen aboard the Kish lightship spotted the drifting, abandoned craft and notified the lifeboat by radio telephone. It was picked up by the lifeboat and towed into Howth Harbour; the voluntary lifeboat crew had been at sea for 11 hours by then.

'He was the bravest little lad I have ever seen,' Kellett said of his nephew afterwards. 'He never uttered a whimper, and when he became so exhausted he lay down in the bottom of the boat and fell asleep.' The following week Michael came fifth of 56 in his class in the children's horse-jumping events at the RDS in Ballsbridge.

A month later Captain Chris Carey recorded his first sea rescue. On 16 September 1964 a fishing vessel ran aground off Caher Island in County Mayo, and the crew of Carey, Sergeant Willie O'Neill and Corporal Alec Dunne rescued one man and two women. However, it would be some time before there was a public awareness about the State's new service. 'No effort was made to advertise the fact that we were there and available,' Commandant Fergus O'Connor says. 'We would open the newspaper on a Monday morning and read about an accident that we might have been able to respond to — only no one thought to call on us.'

And yet it had been a calculated government decision to purchase the new helicopters after the 'big snow' of late 1962 and early 1963. Weather conditions had been particularly harsh during that winter, with heavy snowfalls and severe frost cutting off upland areas for several weeks. The hardship and suffering, particularly among the farming community, caused a public outcry. In one instance, a pregnant woman had to be dragged on a sheet of corrugated iron through the snow to reach hospital in south Wicklow. In the absence of any State-supported air-sea rescue service, members of the Irish Parachute club had valiantly given of their time to bring supplies to people cut off on high ground or on offshore islands in times of bad weather. Irish Parachute club volunteers were often the first to the scene in emergencies where the assistance of British military helicopters had to be called upon.

'Is the machinery of the Department of Finance so creaking that it cannot produce the price of a helicopter tomorrow?' thundered an *Irish Times* editorial on 2 January 1963. 'And is the Department of Defence so entrenched that it cannot direct the Air Corps to crew and maintain such an essential piece of equipment within hours of its delivery? It seems nonsensical that anybody should have to campaign in such a safe cause. Perhaps the snowstorms of recent days, and the fact that there are votes to be obtained in County Wicklow, will suffice to prod the Government and its servants into immediate action. The winter, for all we know, may be only beginning.'

That previous year, 1962, an interdepartmental committee had concluded that the State couldn't afford a helicopter service dedicated solely to search and rescue — although this was disputed by several minority reports within the committee's membership. It was one of two such committees appointed to look at the feasibility of providing helicopters to complement the excellent work of the voluntary Royal National Lifeboat Institution (RNLI), which had first established itself on the Irish coastline in 1803.

As maritime historian Dr John de Courcy Ireland has noted, the Dublin port authority's decision to set up rescue stations around Dublin Bay was designed to protect ships approaching one of the most dangerous stretches of coastline in Europe. It represented the first co-ordinated lifeboat service in Europe. Almost two decades later the Royal National Institution for the Preservation of Life from Shipwreck (RNIPLS) was founded in Britain, and it extended its remit to the Irish east coast where several more stations were founded — although the Dublin Bay lifeboats remained independent of the organisation, which was renamed the RNLI in the 1850s, until the early 1860s.

The turning point was 9 February 1861, when 16 ships and many lives were lost in a terrible storm which swept right up the Irish Sea. In 1862 the three stations at Howth, and Dun Laoghaire and the Pigeon House on the south side of the Liffey were transferred from the Dublin Ballast Board to the RNLI, and the board agreed to give the sum of £50 a year towards maintaining them. The volunteer crews at these stations often found themselves putting to sea in the most dreadful weather conditions, in 35 foot rowing boats fitted with sail which were designed for self-righting in a capsize.

One of the most extraordinary lifeboat rescues during those early years was the call-out to an Australian brig, the *Tergiste*, which ran up ashore in an easterly gale on rocks between Lambay Island and Howth in north Dublin in November 1859. The Skerries lifeboat was hauled over on its carriage to the strand south of the village of Rush and launched with 13 crew under the command of coxswain Joseph Clarke. The station's honorary secretary, Henry Hamilton of Balbriggan, was also on board. However, it couldn't reach the brig and returned to Rogerstown where the lifeboat was kept moored in the estuary.

The brig managed to survive that night. Another attempted rescue by a steamer on 16 November failed due to the conditions. Early on 17 November the wind eased somewhat and the lifeboat crew tried again. After over two hours of constant rowing, they managed to reach the brig and took its complement in twos and threes over the ship's stern. The ship's hull was subsequently salvaged and taken in tow to Dun Laoghaire. Henry Hamilton received the RNLI gold medal for the rescue — one of many to be given to RNLI crews over the following decades.[1]

Throughout those decades there were many acts of heroism, and there were also fatalities among lifeboat volunteers. One of the worst was in 1895 when the Dun Laoghaire number two lifeboat, *Civil Service No. 7*, was launched to assist a Finnish steamship, the *Palme*, which was dragging its anchors off Dun Laoghaire in a strong gale. The lifeboat capsized some 600 yards from the steamship, failing to right itself in the heavy seas. All 15 on board were lost, while the crew of the *Palme* watched on helplessly.

The Dun Laoghaire number one lifeboat, *Hannah Pickard*, then launched with a crew of nine who were assisted by six volunteers from a British military ship, HMS *Melampus*. It also capsized but managed to right itself; however, it had lost its mizen-sail and some of its oars and was forced to return to shore at Blackrock. The Poolbeg lifeboat, the *Aaron Stark Symes*, was launched but was unable to reach the ship.

Next morning, Christmas Day, the 20 crew on the *Palme* were rescued by a lifeboat lowered from the steamship, *Tearaght*, under the command of Captain Thomas McCombie. It was a perilous attempt in continuing heavy seas, but it was successful and McCombie was awarded the gold medal by the RNLI. Back in Dun Laoghaire a telegram was sent from the Queen of England extending her sympathies to the bereaved families of the 15 lifeboat crew.

A public funeral was held for 13 of the 15 bodies recovered, and the sum of £17,000 was raised locally for the families which included a contribution from the RNLI. 'By melancholy appropriateness, the crew was divided between the creeds that are said to divide our country in the same proportion as in the island of Ireland today', Dr John de Courcy Ireland wrote.[2] Every Christmas the Dun Laoghaire station marks the fatality with a wreath-laying ceremony at sea; a granite stone in the harbour records the valiant contribution of those on the last mission by *Civil Service No. 7*.

During the early years of the twentieth century there was a revolution in lifeboat design when engines were fitted for the first time. Not only did this extend their range but it also provided essential speed and safety. The Irish east coast's first motor lifeboat was deployed at Wicklow in 1911. Arklow, Co. Wicklow, Dun Laoghaire, Wexford (which later moved to Rosslare), Howth, Clogherhead, Co. Louth, and Poolbeg in Dublin followed.

Naturally there was some rationalisation of stations, given the greater range. In 1927 Rosslare Harbour's lifeboat became the first to be fitted with a wireless set, which was housed in a watertight casing and promised a communication range of over 80 miles. Rosslare's selection was deliberate, for the ferry port stood watch over some of the most treacherous and exposed seas on this coastline where many ships had foundered over the centuries; and over the years its crews and fellow volunteers in the neighbouring fishing port of Kilmore Quay, Co. Wexford, were to distinguish themselves on many occasions.

One of the first such acts of heroism was at Kilmore Quay nine years after the establishment of its first lifeboat station. An RNLI silver medal was awarded to Dennis Donovan, chief boatman on the 34 foot ten-oared *John Robert* for the rescue of five of the crew of a brig, the *Isabella*, which had run up on rocks in a storm.

Then there was the memorable morning of 27 November 1954, when the Liberian tanker *World Concord*, *en route* to Syria from Liverpool, broke in two during violent storms in the Irish Sea. Waves were reportedly reaching 20 feet when the Welsh lifeboat from St David's reached the ship's stern and took 35 crew off in 34 hazardous approaches. However, there were still seven crew on the bow section of the ship and the Rosslare lifeboat, the *Douglas Hyde*, put to sea to try and locate it some 28 miles east-south-east of the harbour. Conditions were said to be so bad that the mailboat running between Fishguard and Rosslare had taken six hours to make the normal three-hour passage.[3]

A Royal Navy destroyer, HMS *Illustrious*, located the bow of the tanker just after 7 p.m. and informed the lifeboat. When coxswain Richard Walsh and his crew reached it, it was drifting northwards at a rate of about three and a half knots, but the seven on board didn't seem to be in immediate danger. Walsh took a calculated risk: it would be far safer to attempt an evacuation during daylight. Throughout that night the lifeboat stood by the tanker's shell in horrible seas, with west-south-west winds reaching storm force by midnight.

By dawn the tanker's damaged bow was listing to port and seas were reaching 25 feet in a heavy swell. In a highly hazardous manoeuvre, where a constant eye had to be kept on protruding parts of the tanker's superstructure, the lifeboat came alongside a 25 foot ladder on the tanker's hull. It took 15 minutes to take the seven survivors off, and every second counted. When the Rosslare lifeboat headed for Holyhead with the survivors — for it was now closer to the Welsh coast — it had been at sea for 26 hours.

Not surprisingly, coxswain Walsh was awarded the silver medal, and bronze medals were given to second coxswain Billy Duggan and mechanic Dick Hickey. The RNLI's gratitude inscribed on vellum manuscript was presented to assistant motor mechanic John Wickham, bowman James Walsh and lifeboatmen Richard Duggan and John Duggan.

The coxswain of the St David's lifeboat, William W. Williams, received a silver medal, and the rest of his crew were also decorated. The *World Concord* rescue was a singular example of endurance and one which inspired a ballad, and a letter from the late Greek shipping magnate, Aristotle Onassis, who expressed his gratitude to the crew by sending them £5 each.

It didn't break the record set by the Ballycotton lifeboat off the east Cork coast in 1936, when it was launched to rescue the crew of the *Daunt Rock* light vessel. The volunteers on Ballycotton's *Mary Stanford* were awarded a gold, a silver and four bronze medals, and the lifeboat itself was also given a gold medal for an extraordinary mission lasting over 63 hours (the vessel was absent from the station for 76 hours and 30 minutes in total).

Major developments in lifeboat rescue took place in the 1960s when inshore lifeboats were introduced — the first on Ireland's east coast being at Howth. Also, a US Coastguard lifeboat purchased by the institution was adapted for a new building programme; this Waveney class vessel was self-righting and had greater speed than the existing nine-knot vessels. The first of its type, the *John F. Kennedy*, was stationed by the RNLI in Ireland at Dun Laoghaire in May 1967, almost four years after the initiation of the State's first air-sea rescue service. By then Dun Laoghaire lifeboat station had become the first to carry out official training exercises with the Air Corps, and by a happy coincidence the then honorary secretary, Dr John de Courcy Ireland, was 'volunteered' to be the first 'victim' requiring an airlift from the lifeboat deck. 'I remember that the Inspector of Lifeboats told us to go way out off the coast, so that if anything happened the public wouldn't see!' Dr de Courcy Ireland says. 'Just as I was dangling in mid-air the mailboat passed, and I think every passenger with a camera took a photograph!'

Throughout the 1960s and 70s offshore lifeboat design was further refined, with one of the contributions coming from the distinguished Irish naval architect, the late Jack Tyrrell of Arklow. When the 52 foot Arun class was introduced on this coastline in the early 1980s, it was capable of speeds of between 18 and 20 knots and a range of 230 nautical miles, while the Trent and Severn classes could extend to 25 knots with a range of 250 miles.

In spite of improved safety, however, there were always going to be risks. On 24 December 1977 the RNLI's Irish service recorded another tragedy. A red flare had been spotted at sea, and the Kilmore Quay Oakley class lifeboat, the *Lady Murphy*, was launched from the tidal station. Coxswain on the crew of seven was Thomas Walsh, and a gale was blowing south-westerly. As there were no lights on the pier, people had driven down and switched on the headlamps of their cars to assist the lifeboat.

The *Lady Murphy* was only half a mile from shore when it rolled over. The crew didn't have their harnesses clipped on deck the first time, but all were recovered. There being no sign of any vessel in trouble, the lifeboat had turned back and was within 200 to 300 yards of the pier when it capsized again. Somehow, all the crew were hauled back on board, except for one, Fintan Sinnott, who had slipped away. There were frantic efforts to find him, but in the black night and boiling seas there was no sign of him.

The following morning the crew put to sea again at dawn to search for his body. They were heartbroken for it appeared as if the call-out the night before had been a false alarm — or a cruel hoax. Coxswain Thomas Walsh and motor mechanic John Devereux were decorated by the RNLI for their courage on that dreadful night, and the crew received the RNLI's thanks. A special vellum was awarded posthumously to Fintan Sinnott. It was the first time that an Oakley had capsized on service, and the last recorded loss to date of an RNLI crew member on the Irish coastline.

THE LIFEBOAT SERVICE HAD already proved its worth, and its vessels were to become even faster and safer in subsequent years when the government decided to establish an air-sea rescue service in 1962. Several serious incidents at sea had concentrated official minds. Five people drowned in Clew Bay, Co. Mayo, on 22 October 1957 while *en route* from Clare Island to the mainland at Roonagh pier, near Louisburgh. In October and December of 1961 there were two serious shipping incidents off Eagle Island, Co. Mayo, and the Wexford coast respectively. Though lives were lost, several lives were also saved by the RAF which had dispatched its helicopters on both occasions.[4]

This dependence on British rescue services was to extend well beyond the initiation of the new service at Baldonnel in late 1963; but even the RAF had taken some time to accept the need for dedicated air-sea search and rescue. In fact it was its marine counterpart, the Royal Navy, which pioneered the service through its fleet air arm after the Second World War.

During the 1940 Battle of Britain the high number of British fighter pilots dying within sight of the cliffs of Dover had prompted questions at the highest level. The same pilots had a far better chance of surviving if their planes were shot down on the French side of the Channel, due to the more advanced stage of German rescue services. Unfortunately it also meant that they became prisoners of war.

Early in January 1941 a rescue service was established at the Royal Navy's Coastal Command headquarters — Group Captain Lewis Croke actually coined the phrase 'air-sea rescue'. In the first few months of that year the percentage of ditched British aircrews recovered safely from the sea rose from 20 to 35 per cent. The British Prime Minister, Winston Churchill, took a keen interest in the new service. In a memo which he sent to the director of the new air-sea rescue unit in March 1941, he enquired as to why German fighter aircraft crews were being saved as well. 'They all look the same in the water', came the reply.[5]

The Royal Navy established Britain's first all-helicopter unit in the late 1940s when it acquired Hoverfly and Dragonfly helicopters made by the US company, Sikorsky. The Dragonfly, which was subsequently manufactured by Westland under licence, was used to save many aircrew from ditched aircraft during the Korean War. During the 1940s specialised equipment was developed for sea rescue situations, including the airborne lifeboat kitted out with engine, supplies and radio which was designed by the famous yachtsman, Uffa Fox.

Several ideas were also adapted from the German rescue services, including the use of yellow helmets in the water which could be more easily identified, inflatable one-man dinghies and a series of rafts which were moored along the English coast. Other early developments included an agreement with Trinity House, provider of the lighthouse service and navigational aids, to fit all navigational buoys with ladders in case any ditched pilot was lucky enough to land near one.

RAF pilots had an ambivalent attitude towards the use of helicopters at first, and couldn't quite see that they were real aircraft — or that they had any practical use. In 1951 the British Admiralty announced it was formally adopting helicopters for rescue work, mainly for use with aircraft-carriers. From then on much training time was spent refining winching gear for rescue operations. The RAF deployment of helicopters to Malaya was to prove an invaluable training platform, as was the response to extensive flooding in East Anglia and Holland in 1953.

During the flooding, Dragonflies from the Royal Navy's squadron at Gosport were dispatched to Holland, where the aircraft rescued over 600 men, women and children from the roofs of houses and other buildings during a five-day period. Doors, seats and other fittings were stripped from the helicopters during the rescue flights to allow the carriage of up to five survivors during each mission. The highest individual total rescued by one pilot was 147, recorded by Squadron Leader Kearns of the Navy's fleet air arm, according to Peter Whittle and Michael Borissow, authors of *Angels Without Wings*, a history of Britain's search and rescue squadrons.

Kearns told the authors that the experience taught the crews many things about helicopter rescue techniques, including the introduction of the rescue strop. Hitherto, aircrews had been using a hook which was lowered and attached to loops in life vests — or 'Mae Wests' as they were known. However, civilians in an emergency usually had no Mae Wests, no suitable clothing to hold a hook, and one of the Dutch helicopter pilots designed a canvas strop that casualties could put their arms over, with the help of a winchman.

In 1955 the helicopter became an official part of RAF search and rescue and Number 22 Search and Rescue Squadron of Coastal Command was established at Thorney Island in Hampshire. Its commander was given the key to an empty room and a telephone in a converted hangar. The yachting paradise of the Isle of Wight and the Solent proved to be an invaluable training ground for both RAF and Royal Navy helicopter crews — the latter based 15 miles west of Thorney Island at Lee-on-Solent. Duty crews at the two bases built up a 'friendly rivalry'; whichever pilot gave the earliest estimated time of arrival to the casualty got the job.

Early on it became obvious that regular training over sea, at altitude and at night was essential if crews were to maintain an edge. As Kearns explained to authors Whittle and Borissow:

> Low flying by day is one thing, with a clear horizon and objects to sight on all the time. But at high altitudes there is a feeling of being suspended in a rather flimsy glass cabin with nothing underneath you and vertigo is a very real danger.
>
> The engine sounds different, the rotor blades don't bite as they do near the ground and you have nothing to sight on. It is easy to feel unsure unless you do it regularly enough to have absolute confidence in your instruments and your machine.
>
> Flying over the sea can produce the same sort of problems. On a grey day in the North Sea, the sea and sky merge together, the horizon disappears completely and it is very easy to be much lower than you imagine.
>
> So it is just as important for the pilots to fly regularly in all conditions as it is for the crew to practise winching and other emergency drills.[6]

In the spring of 1964 the third (A197) of the fleet of Alouette helicopters purchased for the Air Corps touched down at Baldonnel. By 1972 there was a fourth, and four more were on order and due to arrive in 1974. By then there was a set pattern of training which would be further developed when the Air Corps linked with other bodies and agencies. A 'body scoop' made by the Air Corps was first used on a training sortie in September 1967.

And if officialdom in the guise of the Department of Defence proved to be frustrating at times, many individuals around the coastline gave invaluable assistance to the aircrews during those early years. One such was 'Skipper Pat' Griffin in Skerries, Co. Dublin, who had agreed to work with the helicopters on training. He never refused to launch his vessel when asked, and in really rough weather during late 1966 he was often seen to lash himself to the wheel.

Though established for the primary role of air-sea rescue, the crews found themselves responding to all sorts of missions in those early years, and they weren't all offshore. One such was on 16 November 1966 when constant heavy rain caused flooding. In Tipperary the River Suir burst its banks and several farmers who were out trying to rescue sheep got caught in high water.

Army and Civil Defence personnel set out in a small boat to rescue the farmers, but two of them got caught themselves in the current and had to shelter in a tree. The Alouette III was called out, flown by Lieutenant Fergus O'Connor, with winch operator Corporal John Joyce and winchman Corporal Alec Dunne on board.

'When we located all four men, we tried to lower Alec down through the branches but the cable wasn't long enough,' O'Connor recalls. 'So we flew upstream, dropped Alec in the river and he used the current to bring him down to the stranded men. One of them panicked and wouldn't leave, and so Alec had to give him a bit of a clip to get him moving!'

The four men were rescued by the Alouette in 20 minutes, and Alec Dunne was nominated for a Distinguished Service Medal by the military for his heroic efforts. His pilot, Fergus O'Connor, remembers how Alec lost his helmet during the rescue. 'It was almost fully dark when we later spotted the helmet in the swirling current while returning to an overnight base in Clonmel Army barracks. Alec insisted on going back down on the wire to retrieve it from the river — which he did.'

3

CLIFF-FACE CHALLENGES

It was a white bedsheet, trying its best to fly in the Atlantic wind, and Lieutenant Fergus O'Connor was more than worried. If it came loose in the helicopter's down-draught and snagged in the rotors, he would be in serious trouble — as would his crew. The Alouette III was on an air-ambulance mission to the Aran island of Inis Meáin and the pilot had been told that the bed linen would signal his destination.

Fortunately the 'flag' was kept under control as the helicopter approached to land on a piece of level ground running up from the island's eastern shore. However, as he descended, O'Connor realised that the field was too small, as were all the fields around it. The design was a deliberate one by centuries of island farmers who used dry stone walls to break the wind.

The pilot signalled to several men close to the house who were watching him from below. Without any hesitation five of them walked over and pushed the dividing wall down; it might have been standing there for over 100 years. The quick demolition was enough to give the Alouette room to manoeuvre. It would probably take many hours and much sweat to rebuild the wall again.

The pioneers of the new air-sea rescue service, Barney McMahon, J. P. Kelly, Fergus O'Connor and Chris Carey, were on a constant learning curve, as were the ten non-commissioned officers who comprised the winching crew and who also provided the vital technical services on the ground. A high percentage of the early work involved air-ambulance missions and mountain rescue, which meant that winching crews had to undertake medical training. Consequently close links were forged with the Association for Adventure Sports (AFAS) and mountain rescue teams, and in 1973 several aircrew attended one of the first of many mountain rescue courses at the National Adventure Centre in Tiglin, Co. Wicklow.

Links were also established with specialist medical personnel, most notably the rehabilitation unit in Our Lady of Lourdes Hospital in Dun Laoghaire, Co. Dublin, where accident victims with suspected spinal injuries were treated. When injuries such as these became more frequent as motor traffic increased on the roads, Ireland pioneered the use of a flying team to go to the site of an accident, according to Dr Thomas Gregg of the National Rehabilitation Centre. Writing in a special edition of *An Cosantóir* in March 1985, he noted that co-operation between spinal unit medical staff and Air Corps crew had saved many accident victims from permanent paralysis over the years. The hospital's medical staff also designed a spinal stretcher with a traction facility for use on the helicopters.

The first air-ambulance mission on record was to Our Lady's Hospital in February 1964, when there had been a heavy snowfall along the east coast. The Alouette III A196 was fitted with snow skis to take a patient from Wexford up to Dun Laoghaire for specialist treatment. Another early medical flight or 'casevac' required some diplomatic clearance when a seriously ill Cork woman, Maureen Collins, was flown from hospital in Cork to the Royal Victoria Hospital in Belfast on 10 September 1965. The journey was done in two legs by the Alouettes — the first to Baldonnel by Commandant Barney McMahon, and the second up north by Lieutenant Carey.

Carey was told in his briefing that the Royal Victoria was on the right side at the end of the motorway into Belfast. In fact it was on the left. He landed the Alouette in a flowerbed in Musgrave Park Hospital and remembers he had to 'depart rapidly' when accosted by a 'formidable matron'. After he landed successfully at the Royal Victoria, he was asked by Northern reporters if he had got lost. He denied this 'vehemently', he says.

There were also both ambulance and weather relief runs to the offshore islands, and it was here that crews experienced some of the warmest receptions on arrival. 'Of course many of these runs were emergencies, but when you had the time there was always the offer of tea and a little *buidéal* of poitín,' McMahon remembers. A sense of humour was vital on several occasions: McMahon was called out to one of the islands to fly a pregnant woman to hospital, but the nurse attending her in the house couldn't get her into the Air Corps stretcher. He ventured to assist but was banished with a loud '*Amach!*' from the patient's bed.

Three fishermen owed their lives to the Air Corps in June 1967, when they were rescued from their vessel off Easkey, Co. Sligo. The three, Patrick Lynott, Tom McHale and Brendan Brady, all of County Mayo, had been fishing in the 26 foot *Cathal Brugha* a mile off

Lacken pier when they experienced engine failure. The alarm was raised after the boat failed to return ashore that evening.

The Alouette crew spotted the three at about 2.30 p.m. on 20 June 1967 when their vessel was close to rocks in a rough sea. Skipper Lynott, who was reported to have survived the Cleggan Bay disaster which claimed the lives of 45 fishermen from Lacken, Inishbofin, the Inishkea Islands and Cleggan in 1927, said afterwards that he was never worried. 'We kept to the deep water where we dropped anchor,' he told *The Irish Times*. 'Our spirits were high at all times because we knew we would be picked up.' The men had eaten a sandwich each during the night and chewed tobacco when they ran out of matches.[1]

Two days later the Air Corps was involved in a very personal tragedy when three fellow pilots with the State airline, Aer Lingus, were killed in an air crash in County Meath. The Aer Lingus Viscount, *St Cathal*, was on a training flight when it crashed and landed upside down in a cornfield about two miles east of Ashbourne early on 22 June 1967. Captain Hugh O'Keeffe (37), married with two children, from Sutton, Co. Dublin, and two Dublin-based cadets, Rory de Paor (20) and John T. G. Kavanagh (20), died in the accident.

An Alouette from Baldonnel flown by Lieutenant Ken Byrne (now Lieutenant Colonel and retired from the Air Corps) was first to locate the wreckage. The telescoped aircraft was still burning, and Byrne flew over to the main road to guide the fire engines and ambulances into the cornfield. It was the second fatal crash in the State airline's history. (On 10 January 1952, a DC3 with 20 passengers and a crew of three crashed into a mountain in Wales *en route* from London to Dublin.)

Just a week after that there was another aircraft accident, this time in County Wicklow. A Cessna single-engine plane with a pilot and four members of a film crew on board left Ballyfree airstrip, Co. Wicklow, to inspect film sites in the west of Ireland on 29 June 1967. It never made it across the Shannon, as it crashed into the summit of the 3,039 foot mountain, Lugnaquilla. The aircraft somersaulted down into a bed of rock, ripping off the propellor, engine and one of the wings. The pilot, Joseph Durnin, sustained minor injuries, but his four passengers, who were in a bad way, were trapped in the wreckage. He made his way to the foot of the mountain and raised the alarm. Lieutenant Chris Carey, who was on duty at Baldonnel, flew one of the two helicopters which was tasked to assist.

Commandant Barney McMahon travelled to the car park in Glenmalure at the foot of the mountain to act as ground co-ordinator.

Carey headed for the summit, but found the top 300 feet to be shrouded in cloud. He was faced with a difficult challenge. Turning the helicopter into the mountain, he ascended at 'walking pace' with the rotor blades only feet away from the rocks. All the while he had to keep sight of the ground and the rock face in front of him. If he lost visual contact for even a second, he risked becoming disoriented — with potentially fatal consequences. 'After two passes I located the crash,' Carey recalls. 'It had flipped over and was lying inverted just short of the summit. One of the passengers was seriously injured and was still strapped in the aircraft upside down. We stretchered this man to the base and returned to search for the aircraft again. After two more passes we found it.' The pilot and crew took the remaining passengers. 'We were running low on fuel. We were marginally above the weight limit, but I needed to get everybody off the mountain then.' The Alouette landed safely in the car park and some 45 minutes later all four survivors were in hospital.

Two months later Carey made the front pages of newspapers when he had to fly an English diver, George Best, from Bantry in west Cork to Derry for urgent decompression treatment. The diver had developed the 'bends' while involved in marine work off Whiddy Island on 29 August 1967. At the time HMS *Sea Eagle* at the NATO base in Derry had the only decompression chamber on the island. The alternative was Portsmouth in southern England, but doctors had advised that it would be too risky to carry Best higher than 1,000 feet in his condition.

Throughout the early years there was close co-operation between the Air Corps and the RAF, which had advised the Defence Forces to buy the Alouette III in the first place. 'If we had problems, we would talk to them in the air during the daily check flights,' McMahon says. 'They were only 60 miles away.' When the first exchange visit was set up with the RAF, one of the trips arranged for the visitors was to Inis Mór in Galway Bay.

Both British and Irish emergency services co-operated closely when Aer Lingus flight EI 172 with 57 passengers and four crew on board fell into the sea near Tuskar Rock, off the County Wexford coast, on 24 March 1968. The Viscount aeroplane had been *en route* from Cork to London, when the pilot made a radio call at 10.58 a.m. and reported 'five thousand feet, descending, spinning rapidly'. It was the worst air accident in Irish aviation history and one which generated much speculation about possible causes over subsequent years.[2]

In the immediate aftermath, however, there was a rescue operation which rapidly transformed into a search when it became obvious there were no survivors. The Rosslare and Kilmore Quay lifeboats brought in the first bodies and wreckage, and Willie Bates, Kilmore Quay skipper of the fishing vessel, the *Glendalough*, was the first to locate the crash site near Tuskar lighthouse three months later. Even the toughest among five voluntary lifeboat crews found the few days after the crash to be a horrific experience.

Barney McMahon flew five searches between 28 and 30 March for a total of 13 hours, north, south and west of the Tuskar light. 'The RAF and the Royal Navy helped out, and if the RAF Nimrod fixed-wing or the Air Corps Dove spotted something, it would drop a flare. The Alouette had to pick up everything we spotted, even dead seagulls covered in oil. Sometimes you'd see something and it would turn out to be a plastic bag dumped in a river somewhere by a farmer.'

On the night of 21 December 1968, five west Cork fishermen set off from Kilmackalogue in County Kerry in the 60 foot vessel, the *Sea Flower*, for Castletownbere fishery harbour after reportedly refusing a lift by car. It was about 6 p.m. and a depression was sweeping in over the Irish west coast. That night, winds were gusting to 75 miles per hour on the south-western seaboard. In fact it was akin to a hurricane when Michael Crowley, a 30-year-old father of two from Castletownbere, and his young crew of Niall Crilly from Cork city, John Michael Sheehan of Dursey Sound, his first cousin, Noel Sheehan from Dursey Island, and Bernard Lynch from Castletownbere, found themselves in serious trouble.

At around midnight a Dutchman saw flares off Kenmare and called the gardaí. The Valentia lifeboat was tasked but couldn't find anything, while gardaí from Castletownbere, Eyeries, Lauragh, Kenmare and Killarney searched the coastline through the night. An Alouette from Baldonnel took off at first light, flown by Lieutenant Fergus O'Connor. 'By the time we could fly it was really too late,' O'Connor says. 'It was yet another indication that night flying capability was essential if more lives were to be saved.'

The wreckage of the *Sea Flower* was found on rocks in the inlet known as the Kenmare River, and debris was scattered over a mile wide. The first bodies were recovered some hours later. 'When I arrived in bright sunlight, even though the wind had abated a little, the Kenmare River and windward shoreline was still totally covered in foam created by the extreme weather conditions,' O'Connor says. 'I haven't seen a sea like it since. All that remained of the *Sea Flower* was her keel high on the rocks, identified by the enmeshed brightly coloured fishing nets, and there was a group of distraught onlookers close by.' His logbook recorded a wind speed of 50 knots with gusts of over 70 knots on that morning flight. On his way over he remembered approaching Dunmanway and watching a Volkswagen 'pass him out' on the road below. 'I was flying 110 knots, but that showed my ground speed against the headwind.' That gale would have abated from the previous night, he points out. The five young fishermen never really had a chance.

O'Connor was involved in another grim mission on the west coast the following year when nine people, most of them young schoolgirls, drowned in New Quay, Co. Clare, on 29 June 1969. The girls were among a party on board the *Redbank,* a 25 foot launch which had only been delivered to the Redbank Oyster Company in County Clare a couple of weeks before. It was built to carry a maximum of three crew, but there were almost 40 on board, enjoying a pleasure trip after the official blessing by the local parish priest, when it capsized.

The search for bodies was extensive, involving the Air Corps, the Aran Island lifeboat, local fishermen, the Civil Defence, and Army members of the Curragh Sub-Aqua Club who were on holiday in the area at the time. O'Connor was pilot on duty when one body was found by the helicopter; he remembers that Alec Dunne lifted it using the aircraft's body scoop.

During McMahon's time the Alouette was so successful that increasing demands were put on aircrews. The helicopters were tasked on fishery patrols in co-operation with the Naval Service. 'The idea was that if we saw foreign trawlers that shouldn't be there, we would fly over and let them know. I remember one crew throwing fish at us in the helicopter. My colleague, RAF Squadron Leader John Weaver, who came over here on the first exchange visit, suggested we equip ourselves with a bucket and catch the missiles on the next occasion!'

McMahon flew actively up to 1970, when he was sent to the RAF staff college near Windsor and was then promoted to various administrative roles — retiring as Brigadier General after 42 years of service in 1989. One of his last flights in the Alouette was an airlift of an injured crewman from a Canadian destroyer, the *Skeena*, on 3 April 1969. With him were Lieutenant Gabriel Rodgers and Corporal (subsequently Sergeant) Paddy Carey, on a mission which stretched the fuel limits of the helicopter as the distance from Blacksod light in north Mayo was 70 nautical miles. Fortunately air cover was provided by an RAF Nimrod and the patient was delivered safely to the hospital in Castlebar.

The military were reluctant to recognise the efforts of the Air Corps search and rescue teams at first, even though they were often described in newspapers as 'Army helicopters'. It was only in 1971 that the first of several awards was given — three of them for cliff face rescues. Distinguished Service Medals (DSMs) were the highest awards that could be conferred by the military in peacetime. Not surprisingly, the first involved the fearless Barney McMahon. A 19-year-old English student, Gene Gillougly from Teeside in Middlesbrough, who was over in Ireland to attend an interview for a job as a social worker, had been staying with his brother, a medical student at Trinity College. On 18 March 1970 he set off on a day trip to County Wicklow. He was climbing a 200 foot cliff face at St Kevin's Bed with his sister-in-law, Mrs Elizabeth Gillougly, when he slipped and fell about 50 feet and broke his leg.

With weather conditions breaking around him, he clung to a holly bush for five hours, whilst his sister-in-law scrambled down the cliff to raise the alarm. Elizabeth Gillougly was breathless and terrified when she reached Glendalough village, a mile and a half away, and notified a local hotel manager, John Casey. He called the gardaí and the fire brigades at Rathdrum and Wicklow. Bray Civil Defence and the Dun Laoghaire Mountaineering club were called out and made an attempt to reach the student from the west face of the cliff.

The weather continued to deteriorate, with snow, sleet, rain and high winds, and the Air Corps helicopter was called. It made four attempts to reach the student on a most difficult approach. 'For seven tense minutes the three-man rescue team put their lives in danger as the whirling blades came within two feet of the cliff face,' Kevin O'Connor reported in the *Irish Independent*.[3] 'And for a few seconds the upper lake of Glendalough, whipped by a near gale force wind, was menacingly close to the crew.'

Airman Michael Brady was lowered down. Gillougly, who was hugging the holly bush for dear life, was unable to stand and could hardly move his hands. Brady lifted him on to the cable and secured

CITY LIBRARY CORK

him tightly but had difficulty getting the student into the helicopter as he was in so much pain. 'It was a hazardous and dangerous rescue and one of the most difficult I ever performed,' McMahon said afterwards. 'I would say the copter blades were only two feet away from the cliff face at one stage of the rescue. We had to contend with a gale force wind and I had to bear in mind that the down-draught caused by the helicopter might whip away the holly bush and cause the student to tumble into the lake.'

'Yesterday's events again demonstrated the vital need for a helicopter service', an editorial in the *Irish Independent* concluded.[4] 'Certainly it costs a lot of money to keep the service in operation, but it has obvious distinction as a life-saver and that, after all, is a paramount consideration. An expansion of the service would be a worthwhile target. Helicopters are being used more and more frequently in rescue operations off the west coast; so much so that the time would seem ripe for a base, say, at Shannon Airport, to cover the western and southern areas.' The leader writer had some foresight, though it would take many, many years to convince the State authorities of the merits of that argument. Two decades later, after a concerted campaign, a west coast helicopter rescue base was established for the first time on the west coast — at Shannon.

Two years later there was another award-winning rescue in Powerscourt, Co. Wicklow, this time by Captain Tom Croke, Corporal Terry Kelly and Corporal John Ring. During the August bank holiday weekend in 1972, two boys climbing up the spectacular waterfall got into difficulty. Stephen Salmon from Worcestershire in England, his younger brother Christopher and Peter Mellotte from Belfast were close to the top when they got stuck on a ledge. Witnesses shouted to them to stay still while the alarm was raised. However, Stephen tried to move and plunged 400 feet to his death — striking Peter Mellotte as he fell.

Croke had to cope with severe turbulence and down-draughts as he tried to maintain a level that would allow the winching crew to reach Peter Mellotte. The crew were forced to unload all unnecessary equipment to make the aircraft lighter, yet the helicopter was still being buffeted about as it moved in over the boy. Within minutes the winchman was down, had the strop around him and both were winched back up into the aircraft. Christopher Salmon was also rescued unhurt, but he and his parents, who had witnessed Stephen's fall, were highly traumatised. All three Air Corps crew received Distinguished Service Medals for their efforts in a highly pressurised situation.

A TASKING IN WICKLOW several years earlier had proved to be equally testing for the Air Corps crew involved. Lieutenant Hugh O'Donnell, who earned his wings in 1966 and joined the helicopter unit the following year, had to negotiate the ledge of Glenmacnass waterfall outside the village of Laragh in County Wicklow to reach a young County Meath man who had sustained head injuries. Joseph Lynch was out with a group of friends on 14 July 1968, some of them German and French, when he ventured into the river and was washed over the edge of the falls. Had he not been caught by the ledge, he would certainly have fallen to his death.

Several of his friends raised the alarm. Brendan White of Rathfarnham, Dublin, was out sightseeing with Jim O'Shaughnessy, also of Rathfarnham, and two Irish-Americans, John Gallagher and his son, when they were asked for their help. Together with a doctor, who was also in the area, they managed to get out on to the ledge and wrap Lynch in a rug. 'The helicopter arrived very quickly, but we thought it wouldn't get near enough,' Brendan told the *Evening Herald*.[5]

The helicopter had to fly towards the face of the waterfall to approach the ledge, about 50 feet from the top of the waterfall. 'I've never seen anything like the way it was manoeuvred in,' White said. Another eyewitness told the newspaper it was the most dramatic and wonderful rescue he had ever seen. 'One wrong move by the helicopter and the young man would have been sent crashing to his death hundreds of feet to the bottom of the waterfall.'[6]

Airman John Kearney was lowered down on to the rock, bearing a stretcher. He managed to strap the injured man in, clip the stretcher on to the hoist and take it up. The aircraft flew directly to Dr Steevens' Hospital in Dublin where the young man was described later as 'fairly comfortable'.

O'Donnell became the fifth Air Corps search and rescue commander, preceded by Barney McMahon, Fergus O'Connor, J. P. Kelly and Ken Byrne. He flew many memorable missions during his career before he left to join Aer Lingus, and Thomas Walsh from Barnatra in County Mayo and four friends were among those who owed their lives to him. Walsh, Edmund Rogers, a Belmullet bank manager, John Connell, Martin Mills and John Joyce were on their way from Bunbeg in County Donegal to Belmullet in County Mayo in a 26 foot boat named the *Bord Fáilte* on 19 May 1970 when the vessel's engine failed about four miles north-east of Downpatrick Head. It was getting dark and all efforts to restart the engine had

HI-LINE TECHNIQUE

The winchman is lowered to the sinking craft. Generally the SAR helicopter would be in the overhead position as shown. It is possible to deploy the hi-line prior to the winchman being lowered, when the crew of the sinking craft are fishermen used to working with ropes and there is good two-way communication.

The winchman is collected on the sinking craft's deck by the boat's crew. Note that the winchman is wearing a bosun's chair harness.

The winchman pays out the hi-line for a single recovery of a survivor to the helicopter, which would normally be in the overhead position for winching down and recovery. In some circumstances the winch operator would direct the pilot away from the overhead position for ease of flying or if safety considerations dictated. These could include the proximity of masts, aerials and wires on the sinking craft, especially in heavy sea conditions.

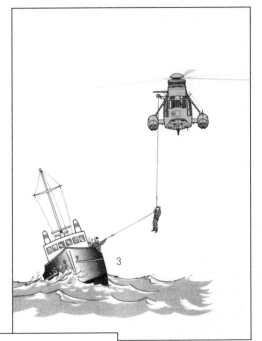

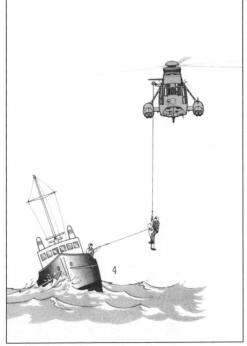

Winchman and survivor are winched aboard the helicopter with a member of the boat's crew paying out the hi-line. Normally the helicopter would be vertically above the fishing vessel.

UP-AND-OVER RECOVERY

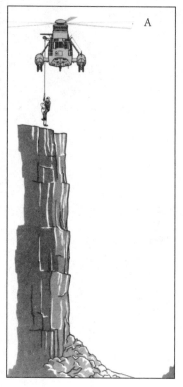

Ideally personnel are recovered into the SAR helicopter with the lowest winch height possible, but where circumstances dictate otherwise the helicopter commences winching at its full rate and climbs.

At the call 'height good', the pilot manoeuvres the helicopter to the cliff top overhead for maximum safety. The helicopter could alternatively move out from the cliff, over the sea, to complete the winching.

Normal recovery is made on the cliff top; alternatively the winchman and survivor are lowered to the surface or a recovery is made via the main cabin door. As usual with all rescue situations it is not possible to generalise.

failed. Fortunately, when they fired several distress flares, these were spotted by several people along the treacherous north Mayo coastline. The Marine Rescue Co-ordination Centre, then run from the Naval Service base at Haulbowline in Cork, was alerted and it tasked the Air Corps helicopter from Baldonnel at 4.20 on the morning of 20 May.

The helicopter was airborne in minutes, with winch crew Sergeant Alec 'Pop' Dunne and Sergeant Paddy Carey on board. The pilot headed out to sea from Ballycastle at around 6 a.m. In the morning light the crew managed to find the drifting boat in heavy seas, perilously close to the rocks off Downpatrick Head. The five men had been at sea for 25 hours when they were winched aboard the Alouette and taken ashore. Even the *Bord Fáilte* managed to make it; it was located and taken under tow into Killala.

Keeping a cool head in a crisis was always a key factor, and this may have made the difference for three more survivors rescued by O'Donnell — this time on 30 March 1973. Three surfers were enjoying the Atlantic breakers at Easkey in County Sligo when they were caught in a strong current and swept towards rocks. Grant Robinson (21) from Enniskillen, Co. Fermanagh, Roderick Allen (20) and David Pearce (25) managed to paddle their boards out to sea to avoid being crushed. However, they were then being dragged further and further out by an ebb tide and an offshore wind.

They waved frantically at the shoreline. By sheer luck, a motorist, Fred Conlon from Sligo, spotted them. He flashed his headlights at them in acknowledgment; though the surfers saw this, they weren't sure if he realised what was happening. However, he did, and he notified the gardaí immediately. A local fishing vessel, the *Sean Foy*, put to sea, but the nearest lifeboat stations were some hours away by sea, on Arranmore Island off north Donegal and on the Aran island of Inis Mór in Galway Bay. The Air Corps was alerted and the crew of Captain Hugh O'Donnell, Corporal David O'Malley and Airman John Ring arrived shortly before 5 p.m. as light was beginning to fade.

With their jet black wetsuits and white boards set against a sea of 'white horses', the surfers were very difficult to spot from the air. O'Malley saw them first, O'Donnell remembers, and he immediately initiated a 'clockface patter' which he relayed to the pilot to keep them in sight.

'The waves were 20 feet high and we were continually being swept off our boards,' Grant Robinson said afterwards.[7] 'If we had lost them we were finished. All the time we were being blown further and further out, and although we were wearing wetsuits and gloves, it was terribly cold. We nearly gave up hope. The helicopter passed over the bay a few times, and once right over us before we were spotted. We were all winched aboard and taken back to Easkey. The airmen did a wonderful job and everyone treated us very kindly.'

The surfers were six miles out and were being carried towards Rosses Point when the helicopter crew found them. The fact that they didn't panic was crucial, Robinson acknowledged. 'We simply concentrated on keeping in touch with our boards, which we knew was the only way we would survive.'[8]

O'Donnell is credited with setting the single-engine helicopter record for offshore Ireland — though he cautions that the RAF may have done some longer missions before that with twin-engine aircraft. An injured seaman had to be airlifted off a Spanish fishing vessel 70 miles north of Valentia, Co. Kerry, in August 1974. In spite of the distance there was 'no debate' about carrying out the mission once the weather was suitable, fuel endurance sufficient, and once an RAF Nimrod was available to provide top cover. The Alouette fuelled at Shannon in transit from Baldonnel and 'topped up' again at Cahirciveen in County Kerry, where some 45 gallon drums had been stored at an Air Corps fuel depot. Flying with O'Donnell were Sergeant Paddy Carey and Corporal (subsequently Sergeant) Dick Murray. Murray's descent on the winch cable was captured on camera by the RAF Nimrod.

The crewman, who had sustained severe limb damage, was winched off using a paraguard stretcher and flown to Tralee Hospital, the round trip taking an hour and 55 minutes — with another 25 minutes to Cahirciveen to refuel. The total flying time was two hours 20 minutes — just ten minutes short of the Alouette III's normal fuel endurance of two hours 30 minutes.

A MAN WHO ALSO kept his head in the most adverse conditions was the late Corporal Brendan McElroy, who was winch operator on a mission flown by pilot J. P. Kelly on 8 March 1970. The Scottish fishing vessel, the *Malcolm Crone* from Aberdeen, had a crew of 12 on board when it lost engine power and found itself drifting towards Melmore Head on the Donegal coast. Winds were blowing 50 knots when the *Malcolm Crone* sent out a distress signal. Another vessel in the area, the *Granton Osprey*, responded and tried to get a line on board. However, though it was eventually secured, because the line was too light the stricken vessel could only be towed at a speed of two knots.

The seas were too rough for the Arranmore lifeboat, and an RAF helicopter at Aldergrove in Belfast wasn't able to assist. At about 4.45 a.m. the Air Corps was alerted at Baldonnel and Captain J. P. Kelly crawled out of bed. 'We flew to Finner camp in an hour and 30 minutes and had to refuel with a hand pump from 55 gallon drums,' Kelly, now Lieutenant Colonel and retired from the Air Corps, remembers. 'We flew in a 60 knot wind through the Gap of Dunloe. The helicopter was travelling at 100 knots, so our ground speed was 40 knots and it took us all of an hour and 15 minutes to get to Fanad. There was a fuel dump at Fort Dunree where we had to refuel again. Our only communication was an aeronautical radio which could contact the radio tower, and a portable radio to contact the rescue units on land. The coaster was very close to the rocks at this stage, and I knew I wouldn't have enough fuel to get the crew off. So I landed at Fanad Head lighthouse, spoke to the skipper of the *Granton Osprey* and said I'd transfer a heavy tow line from one vessel to the other.'

The air crew lowered down their winching hoist and the *Granton Osprey* attached an old hemp rope to it which the winch operator, Brendan McElroy, hauled in. 'I told him not to make it fast in case it snagged and pulled us down, and we tried to move across to the other vessel to transfer it,' Kelly says. 'But the rope began to absorb sea water very quickly, and it got heavier and heavier. McElroy was trying his best to hold on to it, and he had asbestos gloves reinforced with steel. The gloves were burning, so he had to take them off, and the rope began tearing into his hands. As we moved away he eventually had to let go, but at this stage the tears were rolling down his cheeks with the pain.'

A British Army Scout helicopter arrived on the scene at this point, and Kelly asked the pilot to contact the *Granton Osprey* and ask the skipper to drift in on the *Malcolm Crone* to reduce the distance between them. A second line of lighter hemp was transferred between the two fishing vessels, secured successfully, and both got under way and headed for the Scottish port of Ayr. The Alouette had been hovering for 55 minutes by this time, and landed again with very little fuel.

Kelly will never forget his first glimpse of the winch operator's hands. 'Both of his palms and his fingers were absolutely raw, and the skin had just been shredded off. He never got any recognition for that, and the only thanks we got was a rather inaccurate article afterwards in an RNLI publication written by the pilot of the British Army Scout.' Corporal McElroy subsequently left the Air Corps to join Aer Lingus. He died at a relatively young age.

Kelly did his fair share of air-ambulance missions during his time; it was estimated that for every one alert at sea, there were four medical runs onshore or on the islands. One other mission he will not easily forget was from a regional hospital on the west coast. During a routine medical procedure part of a catheter disappeared into a small boy's arm and he had to be transferred immediately for emergency surgery in Dublin. The boy was in good spirits when Kelly arrived at the hospital, and the main challenge was to get him to stay quiet *en route*. There was a serious risk that the catheter would puncture a main artery.

The child was accompanied by a doctor who was able to talk to the pilot on the headset. Within minutes of take-off Kelly felt something on his knee. The little boy had nestled in under his shoulder and was sitting at the controls. 'The doctor signalled to me to do nothing. To upset him at all could have meant very serious trouble. I had kids myself so I didn't mind, and I kept talking to the doctor. But he was 40 minutes on my lap and I was terrified that if we hit any turbulence he could die. He settled in quite happily and seemed more than content. I think he was the only one of us who enjoyed that flight!' The boy had the emergency surgery, and when Kelly enquired afterwards about his condition, he was told the operation was a success.

THE 'DADDY' OF ALL early cliff and mountain rescues, as one pilot described it, took place on 1 August 1977. The pilot, Commandant Paddy O'Shea, actually won two DSMs that year and was the only pilot to hold two such distinctions for the role he played in Glendalough in May and in Donegal three months later.

'There was a group of Northern Ireland teenagers on a cross-border goodwill tour and they were out in Wicklow for the day,' O'Shea says of the first mission, on 13 May 1977. 'Two of the guys got separated, clambered up to St Kevin's Bed and got stuck. I remember we had to go into a hover close to some trees and Airman Blackie (David) Byrne had to make his way through the brambles to reach the ledge.' Byrne had to negotiate an overhanging tree to reach the two youths, both from County Tyrone.

'Blackie had the second survivor on the strop and he was chest to chest with the casualty when his helmet visor got trapped on a branch and he found himself being inverted . . . Luckily the visor snapped and winch operator Dick Murray got them both up.' The two youths were landed at a car park close to the local hotel. The total time on the mission, from Baldonnel to Glendalough and back again, was one hour and five minutes. O'Shea and his two colleagues received second class Distinguished Service Medals.

Even as O'Shea was being nominated for that award, he was called out on a mission which was to earn him a first class medal. On 1 August 1977 several young people attending a function in Falcarragh, Co. Donegal, went out for an evening walk. A Franciscan priest, Father Thomas Rocks, was at Árd Mhuire priory in Creeslough, north County Donegal, when he got word that three people were trapped on Muckish Mountain. He and his colleague, Brother Richard, weren't quite sure whether to believe the news; to their knowledge no one had ever been in trouble on the mountain before.

'It was a bright sunny evening, about 8 p.m., and we drove to within half a mile of the regular climb starting point', he wrote afterwards.[9] 'The only living creatures to be seen for miles around was a flock of mountain sheep. Since it was a beautiful evening, we decided to get out and admire the view which from that point is most spectacular, the vista stretching to Tory Island. The only sound to greet us was the bleating of the sheep and we listened for some time to this lonely and pathetic cry. Then, quite suddenly, mingled with the bleating of the sheep we caught the faint and distant cry, "Help!" Suddenly, the hitherto incredible story became a reality.'

They were on the west foothills of Muckish at the time, and the summit plateau was some 1,700 feet above them. Since they knew it was going to be difficult to work out where the cries were coming from, they decided to follow the path marked out by miners who had worked on the mountain in the 1940s, and as they climbed the voices became clearer. The problem was they still couldn't pinpoint the location. An onshore summer breeze had strengthened and was increasing, 'so that the cries of the trapped seemed to come from one direction, then another'.

'By now it was nine o'clock,' the priest said. 'Dusk was coming on; some vaporous cloud capped the summit; as yet we had not had visible contact with our objective and there was the growing danger that we might not spot them before visibility was reduced to nil. Being attired in habit and sandals, it became almost impossible for me to climb further, so we agreed that Brother Richard, who was more suitably attired, should proceed further up and that I would remain near the foot to liaise with any rescue force that might arrive.'

The two clerics were by then about 1,000 feet up the mountainside and the summit was still another 1,100 feet away. 'Somewhere in that 1,100 feet lay the trapped climbers. As Brother Richard climbed we kept constant vocal contact. We agreed that this was essential, otherwise there was the danger that even if we did spot them, the information would be useless if we lost contact with one another. Finally, after climbing approximately a further 1,000 feet, Brother Richard established visible and vocal contact with the man and young woman. They were stuck on one of the steepest cliff faces of the entire mountain, unable to ascend or descend. I told Brother Richard to remain exactly where he was, to use his handkerchief as a signal, and that I would return to the nearest phone and summon help.'

Father Thomas descended as quickly as he could, scrambling over rocks and gorse, snagging his habit and tearing his sandalled feet. He had just reached the foot of the mountain when an Air Corps helicopter landed beside him. It was an Alouette from Finner camp in the south of the county, and it had been alerted by a third climber with the couple who had managed to make it to the summit and descend by the eastern flanks to raise the alarm.

'It was just possible to see Brother Richard some 1,000 feet above signalling to the crew, and I told them that he could indicate to them exactly where the trapped climbers lay.' The helicopter took off, following Brother Richard's signals, and the crew were able to pinpoint the location of the trapped climbers with searchlights. However, the helicopter wasn't equipped for search and rescue, being based at Finner for military duties and border patrols. It landed and sent a radio message to the Air Corps at Baldonnel to dispatch immediate help.

'It was now approximately 9.30 p.m.' Father Rocks remembered. 'It was no longer dusk, but getting dark. The wind had strengthened further, but thankfully the threatening mist had dispersed. I gathered from the Air Corps helicopter from Finner that landing on the summit was too dangerous, owing to the strength of the wind. So we settled down to await the arrival of the craft from Baldonnel, all the time reassuring the trapped couple vocally that help was on the way.'

At 10.15 p.m., with dark closing in, there was the distant but distinctive sound of a helicopter approaching. The Finner helicopter lost no time in getting airborne again to guide in the Baldonnel search and rescue crew. 'With both craft in the air, their searchlights lighting up the mountain, the roar of engines driving the mountain sheep helter-skelter, their navigation lights winking continuously, the scene was finally set for what came to be described as the most daring and courageous rescue to be carried out by the Air Corps', Father Rocks wrote.

'Having no knowledge of the technicalities of helicopter piloting, I can only describe what I saw, and felt, and feared, as these so versatile craft closed in on the mountain and on the site of the rescue.' The Finner helicopter fixed its landing light on the exact location of the rescue bid and hovered overhead — the priest recognised how difficult this must have been for the pilot, working in pitch darkness and with a strong wind 'bouncing off the cliff face'. Within seconds the Baldonnel helicopter moved in.

'The pilot of the Baldonnel craft had an even trickier task. How close he got to the cliff face I cannot tell, but to us standing 1,200 feet below it looked as if at any moment the blades of his craft must dissolve into atoms by striking the rock face. So immersed were the onlookers in the dreadful prospect of an even worse tragedy that when the craft pulled away from the mountain face, with the first climber safely on board, all breathed a long sigh of relief.'

Rescuing the female climber was most difficult, according to the priest. 'She was closer still to the face of the cliff, wedged in a rock crevice and too scared to move a muscle. Yet the three crewmen made it look simple — pilot, winchman and winch operator. Apparently it was thought advisable not to take the girl on board until the craft was in a safe position, and so we saw the spectacular sight of the helicopter slowly rising above the view lines of the summit, the girl dangling some 20 feet below, and all clearly defined by the unerring aim of the Finner craft. Soon everyone was safely on the ground, the whole operation having taken about 15 minutes.'

What the Franciscan didn't know was that the girl was dangling by accident. O'Shea remembers that the girl was terrified, unable to move, and it was impossible to fly vertically over her as she was in a crevice. 'Sergeant Willie Byrne was winch operator, and Airman Owen Sherry was winchman. Sherry was landed about 25 feet away from her and could only crawl on his hands and knees. He didn't have enough cable to reach the girl, and Byrne couldn't see him, so he signalled to the girl to approach him.'

'Eventually she got into a position where I could grab her by the arm,' Sherry recalled in an RTÉ Radio documentary by Madeleine O'Rourke. 'Unfortunately, just as I grabbed her, the ledge we were both standing on collapsed and we both slid down an embankment for about 25 feet — before going out over a cliff area with about 800 to 900 feet of a sheer drop into the valley.'

O'Shea could feel the aircraft lurch towards the mountain in the black night, rotor blades inches from the rock. 'There was a pendulum effect as they swung below us. I remember I was shining my landing light up through the rotor blades, and Donal Loughnane had his landing lights on the survivors. We turned around and were facing Tory Island and Falcarragh. There were lots of little lights sparkling from the cottages in the village, and for a moment I wasn't sure if that was the sky or the ground.'

Sherry was under severe pressure. 'The only contact I had with the girl at this stage was that I was holding her by the arm and I had been turned upside down on the rescue chair which I was sitting in,' he told RTÉ. 'But fortunately we swung back in over a small grassy ledge, and I had time just as we hit it to put the rescue strop around her before we swung back out again into open air. And all breathed a sigh of relief as we were successfully winched into the helicopter. At this stage the helicopter then moved away from the cliff area and it was decided that we would drop this girl into the car park at the base of the mountain because of the dangers involved in having two casualties on board at the same time, and the weight penalty and that . . . So we moved away and successfully landed at the base of the mountainside, dropped the casualty off and took off once again to fly up to the top of the mountain to effect the rescue of her male counterpart who was some 200 to 300 feet above her.'[10]

The two clerics witnessing the rescue had company by then. A large group of locals had gathered and gave the rescuers and rescued a hearty welcome. 'But the uppermost thought in my mind and in the minds of the many locals who by this time had gathered was admiration and respect for those experts of the air who, with their machines, had possibly saved the lives, or at least averted the real possibility, of two people suffering a night's exposure on the cold and windswept face of Muckish', Father Rocks said.

Two Alouette III crews were awarded the Distinguished Service Medal for the Muckish night rescue: Commandant Paddy O'Shea, Sergeant Willie Byrne, Airman Owen Sherry, Captain Donal Loughnane and Airman Richard (Dick) O'Sullivan.

Father Rocks was left with one abiding thought after his experience. 'I hope our Air Corps always maintains this high standard of efficiency combined with bravery. I hope also that those in authority duly recognise their achievement, and that no cutback in expenditure is ever envisaged in the case of air rescue, but rather that even more money be expended to augment the service. Better equipment is always needed — we have the men.'[11]

4

FROM WHIDDY TO WEST OF FASTNET

The pilot could see the inferno 60 miles away. Commandant Fergus O'Connor was airborne just over Cork Airport when he caught sight of the red glow. It was in the early hours of 8 January 1979, and sea and sky seemed one as fires blazed in Bantry Bay.

At about 12.55 a.m. a 120,000 ton French oil tanker, the *Betelgeuse*, was discharging crude oil alongside a jetty at the oil terminal on Whiddy Island when it was ripped apart in an explosion. A series of blasts demolished the jetty, killing all those on board the ship and on the quayside. As flames threatened to spread across to the terminal, Cork County's special disaster plan was put into effect.

An eyewitness described the devastation to an *Irish Times* reporter for the following day's newspaper. 'It shook the area. I didn't realise what had happened for a minute, then I saw the flames shooting into the sky from Whiddy Island. They leapt even higher as a series of explosions went off. We knew the tanker had gone up. Nobody could have lived through the fire. The jetty was up in flames and the ship was blown apart. . . .'[1]

All available rescue agencies were mustered, including the Garda Siochána, fire brigade units from all over the south-west of the county, medical teams and a fleet of ambulances from local hospitals. Every available craft in Bantry Harbour was requisitioned for emergency transport out to the small island. For a time it seemed as if the island's 61 inhabitants might be in danger as efforts were made to prevent the fire from spreading to the main storage tanks. All were evacuated to shore.

As for the 50 who had sustained the full impact of the blast, including the tanker's entire French crew of 43 and seven local Gulf Oil workers, there was no hope for them. An initial 'rescue' alert relayed to the Air Corps, the Royal Air Force (RAF), the Royal Navy and the Naval Service was to become a search for bodies before the various units had arrived on the scene.

O'Connor wasn't in the Air Corps any more. Having replaced Commandant Barney McMahon as Officer Commanding the Helicopter Squadron, he retired in April 1975 to take up an appointment with Irish Helicopters. On the night of 8 January 1979, he was at home in Cork on 'night emergency standby' for the Kinsale gas field production platforms and was telephoned shortly after midnight about the explosion on Whiddy Island.

He drove immediately to Cork Airport, where the standby Bell 212 helicopter, contracted to Marathon Petroleum, was based. The twin-engine B212 was suitably equipped for instrument flight and night search operations, but not for rescue. Occasionally it had carried out sea searches when requested to do so by the Marine Rescue Co-ordination Centre, then at Shannon. Now its assistance was required, urgently; the Air Corps Alouettes were restricted to daylight operations.

O'Connor landed at the privately owned Bantry airfield in complete darkness, relying on his own local knowledge of the landing area. 'Two Royal Navy Sea King helicopters flew in after me, and there were no lights so I had to marshal them in with torches. I remember that one of the helicopters ignored me and later that morning it emerged that he had landed with the rotor blades just five foot from a pole.'

It was agreed that the helicopters would fly half-hourly sorties along the coast. Already bodies were being recovered from the water by tugs, boats and by divers who had to deal with burning oil. The Naval Service's auxiliary ship, LE *Setanta*, and its two coastal minesweepers, LE *Gráinne* and LE *Fóla*, took turns at spraying the oil as a Navy diving team carried out recovery missions. 'Bodies were brought in and laid out in a hangar at Bantry airfield before being taken to hospital in Cork,' O'Connor recalls. 'I remember walking in and seeing the local priest and a local minister alone in the hangar, holding on to burning stumps of limbs as they prayed. Most of the bodies were covered in oil, badly burned, and it was an experience I will probably never forget.' As a gesture of support, the Royal Navy helicopters flew some of the first bodies to Cork that morning, before refuelling and returning to base at Culdrose in Cornwall.

Dental records had to be used to identify many of those who died. The subsequent inquiry established that there was nothing anyone could have done in the immediate aftermath of the blast to save any of the 50 victims. A further fatality occurred when a Dutch diving operator died during the salvage operation. It was also an accident that should never have happened. A fire had broken out on board the tanker after 80,000 tons of crude oil, two-thirds of its cargo, had been unloaded at the terminal. There had been attempts to extinguish it, and a warning over the ship's radio that the crew should abandon the vessel.

The president of the tanker company, Total, confirmed that the *Betelgeuse* was not fitted with a safety 'inert gas' system to prevent explosions of the combustible gas that can accumulate when oil is being discharged. The safety procedure was not standard when the tanker was built in 1968.

Mick Treacy was an aircraft engineer in the Air Corps when the disaster occurred. He had joined up five years before, in 1974, and had completed two years in the apprentice school before moving into search and rescue. It was what he always wanted to do ever since his first life-saving rescue at the age of just seven when he was walking along the Royal Canal in Cabra. He caught sight of a lady 'dancing' on the side of the canal bank, and ran down to see a toddler floating face down in the water. 'She literally threw me in to rescue the child, which I did,' Treacy says. The toddler was 18 months old and she survived. Several years later Treacy saved two children from a house fire. Uncannily, his son Raymond saved the life of his friend in the Royal Canal — just 100 metres from his father's own rescue effort — many years later. Raymond ran up, stopped a car and three Canadians jumped in and got the boy. 'I was so delighted he did the right thing and didn't try to jump in himself,' Mick Treacy says. A keen mountaineer, Treacy has immersed his kids in adventure sports — and in the safety culture that goes with it.

At the time of the *Betelgeuse*, Treacy travelled down to Whiddy by road to assist in servicing the two Air Corps helicopters dispatched to west Cork. It was his first introduction to the 'deep end', as he puts it. 'Our job was to help clean the inside of the aircraft which was covered in crude oil each time it came in. You didn't really get to see the bodies because they were encapsulated in oil which was quite thick.'

It was to be one of several shocking experiences for him that year. He was also involved in the aftermath of the IRA atrocity in Mullaghmore, Co. Sligo, on 27 August 1979, when four people including Lord Louis Mountbatten, the Queen's second cousin, were killed by a bomb which exploded while they were out on Mountbatten's boat, *Shadow V*.

Treacy was on duty when the scramble bell went off at about 1.05 p.m. 'Captain Donal Loughnane and Sergeant Paddy Carey were on with me. I was winchman. We got a very confused message about something happening in Mullaghmore in Sligo. It was only when we were airborne that we got some more information via radio *en route* and learned that an explosion had occurred.'

The 50 lb bomb had been placed in a lobster pot, which was set off as the pot was being hauled. 'We arrived at the scene and conducted a search, and of course it was daylight and an azure blue clear sea. I was winched into the water to the first of the bodies that we saw. I was lifted again to another one and it was a child — Paul Maxwell, a 15-year-old lad from Enniskillen. Because of the injuries he had sustained, I couldn't put a strop on him and so I held him between my arms and legs.

Lord Mountbatten's body was one of the last to be found. 'My guess is that the kids were out on deck hauling in the lobster pots, and he may have been in the wheelhouse. Substantial parts of the vessel were under water. There were splinters of timber scattered everywhere — a massive amount of debris. After that, it was a matter of searching, locating and trying to avoid further incidents with craft below us.'

The town of Sligo and the community in the seaside resort of Mullaghmore was stunned, and it would be many years before the village would fully recover from the shock. 'Lord Louis', as he was known locally, owned Classiebawn, a castle commanding an unrivalled and exposed view of Sligo Bay, and he had spent many summers in the area with his family. There had been seven people on board *Shadow V* that morning with the Earl — his 14-year-old twin grandsons Nicholas and Timothy Knatchbull and their parents, the boys' grandmother, the Dowager Lady Brabourne who died in hospital from injuries sustained in the blast, and 15-year-old Paul Maxwell who was crewing the angling boat.

The Knatchbulls and one of their sons, Timothy, survived with serious injuries, but Nicholas died. The bodies of the four victims were flown from Sligo Airport by a British military Hercules aircraft to Britain. It was a particularly bleak period in the Northern Ireland conflict. On the same day 18 British soldiers were killed in an ambush near Carlingford Lough on the north-east coastline. The Air

Corps offered its assistance in the emergency response; for political reasons it was decided it should not fly across the border.

The Mountbatten alert came just a fortnight after another major emergency at sea, and one which involved the resources of both the British and Irish military. On 13 August 1979, Lieutenant Commander John Kavanagh, captain of the Naval ship LE *Deirdre* and later to become Flag Officer of the Naval Service, had set out on a routine patrol of the south-west and west fishery grounds. The *Deirdre* — now decommissioned — was then a relatively new vessel, built seven years before in Cork's Verolme dockyard. Kavanagh was about 30 miles west of Fastnet Rock off the south-west coast when he noticed something unusual about the barograph trace, measuring atmospheric pressure, in his cabin. It had taken a dramatic dive.

He checked it, and it was working. There had been no severe weather alert. The BBC radio shipping forecast for sea area Fastnet at 1.55 p.m. that day gave south-westerly winds, force four to five increasing to force six to seven for a time. As the ship headed out west of Fastnet Rock in the late afternoon, the wind had backed from light south-westerly to southerly force six.

The *Deirdre* was 40 miles west of Fastnet, and it was 10 p.m. when the barometric pressure 'really began to plummet', Kavanagh recalls. The wind had increased to southerly force seven with a rough sea and a heavy swell, but it was still 'nothing unusual for the *Deirdre* to contend with', and it appeared to correspond with the conditions forecast. But by 10.25 p.m. the barograph had dropped from 1014 hector-pascals (HPs) at midday to 989. This represented a drop of 25 HPs in ten and a half hours.

At 10.45 p.m. the BBC issued a new gale warning for Fastnet, which indicated that a strong south-westerly gale force nine to storm force ten was imminent. The ship altered course back towards Fastnet Rock. By midnight the wind was strong gale force nine, the pressure had dropped further to 984 HPs, and at 1.40 a.m., just over three miles south-west of the rock, the patrol ship received its first call for assistance. This was one of a series of Mayday calls picked up by the coast radio station at Valentia in County Kerry, and there would be a very long night ahead.

The international Fastnet yacht race, which was part of the Admiral's Cup series, had been hit by a freak summer storm. Some 54 vessels were competing in the 605-mile race from the Isle of Wight, round the Fastnet and back to Plymouth. At that time they were scattered at various points between the Scilly Isles and Fastnet Rock, and they were accompanied by a loose flotilla of supporters and observers. There were over 300 pleasure craft in all with combined crews of about 2,700 men and women on the 600-mile stretch of water. Many were unprepared for what was to hit them, without VHF radios on board to issue distress messages.

The rising winds whipped up mountainous seas which were already foaming and crashing around the granite lighthouse. The Royal Air Force and the Royal Navy were tasked, along with RNLI lifeboats, and assistance was sought from all other vessels in the immediate area, including the Irish Continental Line ferry, the *Saint Killian*. The Air Corps Alouettes were sent to Cork, but due to weather conditions and range the mission was beyond their capabilities.

The RNLI lifeboats from Baltimore, Ballycotton and Courtmacsherry in County Cork and Dunmore East, Co. Waterford, spent 75 hours at sea in the 60 knot winds, which only abated on the morning of 15 August. Culdrose in Cornwall became the communications centre for the rescue operation, which involved seven British military helicopters, three RAF Nimrod aircraft, six Royal Navy ships, a Dutch frigate, the four lifeboats, the Irish Continental Line ferry and the Naval Service patrol ship. The *Deirdre* played a key role from the early hours of 14 August, serving as a communication link between Valentia coast radio, the lifeboats and the Dutch naval frigate, the *Overijssel*, which had been assigned as guardship for the race.

The Navy ship's first distress call had been picked up from a Cork yacht, the *Regardless*, owned by businessman Ken Rohan. It had lost its rudder and was in danger of capsizing, and so the *Deirdre* stood by it about four miles off the Fastnet until the Baltimore lifeboat arrived under the command of coxswain Christy Collins. When a tow was eventually secured after five attempts, the *Deirdre* steamed on to assist other yachts in trouble.

'By this time the wind had begun to veer and was now blowing from a south-south-westerly direction, strong gale nine gusting frequently to storm force ten,' Kavanagh recorded. The *Deirdre's* decklog entry recorded '77' for sea and swell on the Douglas scale, which is akin to seas of between 24 and 36 feet. Barometric pressure continued to fall, reaching 980 HPs at 4 a.m. and then started to rise sharply. Over the next 14 hours the wind was a steady gale force eight, frequently reaching strong gale force nine with some severe gusts and veering gradually to a westerly direction.

Some of the yachts didn't require help; others needed tows. 'The only way of communicating with some of these vessels was to go up very close, sound the siren and wait for an indication by hand signal from the crew as to their status,' Kavanagh says. Between 10 a.m. and noon the ship gradually worked its way eastwards, rolling heavily, with seas rising 30 feet. At about 4.30 p.m. the ship came across a yacht, the *Hestrul II*, some 75 miles south-east of Cork Harbour. There was no sign of life on board. When the ship called up a Royal Navy helicopter in the area, it confirmed that the crew of six had been airlifted off.

At 7.30 p.m. the ship received a radio message from the *Silver Apple* requesting assistance. It had lost its steering and winds were westerly force seven with a very rough and distorted swell. It was around 1 a.m. the following morning when the *Deirdre* located the yacht 37 miles south-south-east of Cork Harbour. It had lost its rudder but had managed to make a jury steering rig and was motoring back into Cork. The ship accompanied the vessel in what was now a light south-westerly breeze, and then headed back out to look for more stragglers.

By now the after deck of the ship resembled 'a boatyard', Kavanagh says. All the rudders which the ship had picked up in the previous two days were made of carbon fibre, a material which just didn't measure up to the severe 'torque and stresses' of the storm. It was about 3 p.m. on Wednesday, 15 August, that the Navy crew spotted a lifejacket, 60 miles south-south-east of Cork, bearing the name *Alvena*. The ship took the vessel under tow into Cork Harbour. The ship's commander learned later that the *Alvena*, a French 33 foot vessel, had been rolled over, dismasted, and the windows were smashed in by the seas. The crew of six took to their liferaft and were rescued by a yacht, the *Moonstone*. At that stage the six had experienced two capsizes in the raft. The *Moonstone*, with 13 on board, made its way back to Falmouth.

For the helicopter pilots and the ships at sea, trying to find small craft in such conditions was extremely difficult. Some yachts had battened down the hatches and were riding out the storm 'under bare poles . . . drifting in the high seas and heavy swell prevailing' as the *Deirdre*'s log recorded. A total of 136 people were saved; 15 lives were lost, including that of Irish yachtsman, 35-year-old Gerard Winks, and it was something of a miracle that the death toll wasn't higher. Eighty-five boats actually finished the race.

Among the yachts hit was the *Morning Cloud*, owned by former British prime minister, Ted Heath. Fortunately for him, his rudderless vessel was taken under tow to Falmouth. Arthur Moss, on board the British yacht, the *Camargue*, explained afterwards that it was not so much the winds as the mountainous seas. 'The boat turned turtle a couple of times and then righted itself again,' he told reporters. 'I think it was sheer wave power that did it. The waves were simply enormous.' Several crews praised the heroism of the Baltimore lifeboat; Nick Clayton, on board the *Regardless*, couldn't thank them highly enough. 'Anyone else would have said "That's it, you'll have to leave her." But they kept coming back again and again.'[2]

The *Deirdre* continued searching until the evening of 16 August, returned to the Naval base and then sailed again the following morning. Early that afternoon of 17 August it recovered yet another carbon-fibre rudder; shortly afterwards it came across the upturned hull of the *Bucks Fizz*. The 38 foot trimaran wasn't actually competing in the Fastnet, but was one of the tragic casualties. When spotted by a rescue helicopter 40 miles south of Cork Harbour on 15 August, a winchman was sent down. He found nobody on board. Two bodies were later recovered by a passing US freighter, and the remaining two crew were never found. Among the crew had been a man and a woman who had only met on the eve of the race in a pub in Cowes and had been invited along for a sail by the trimaran's owner. The hull was very badly damaged and proved impossible to tow, Kavanagh says. As it was a risk to navigation, the ship's crew dismantled it, took the individual hulls on board and returned with it to Cork.

Back 'again and again' to the Irish coastline came British rescue helicopters throughout the 1980s, when missions proved to be beyond the range of heroic lifeboat volunteers and the sturdy Alouette III fleet. In 1981 an Aerospatiale SA 330 Puma medium-range helicopter was leased by the Defence Forces at a time of increased security commitments. The first twin-engine helicopter to be based at Baldonnel, it had been custom built by the French military as a troop carrier and played a key role in joint Army Ranger/Air Corps operations. It was acquired to provide the Air Corps with basic experience in flying a multi-engined, medium-lift/range helicopter before a final decision was made on purchasing new aircraft.

Early in 1982 the Puma was to prove its worth in search and

rescue also when severe snow hit the eastern half of the island and cut off many hill and mountain communities for days. Over a ten-day period the Puma and the Alouettes carried out 148 snow relief missions, saved or assisted 98 people and airlifted livestock stranded on hillsides in the blizzards. Flying and technical crews worked around the clock, and for this the squadron was later presented with the Fitzmaurice Award for Services to Aviation by the Irish Airline Pilots' Association.

The following year the Puma lease was terminated, and the helicopter was flown back to its owners in Marignane, France, in February 1983. The decision, which was later criticised in several government reports, was a severe blow to morale and one which was to mark a turning point in Air Corps search and rescue. It seemed to put paid to all hopes of developing medium-range rescue capabilities and reducing reliance on British air-sea rescue units.

However, even several Pumas would have made little difference to the survival chances of 329 passengers travelling on board an Air India aircraft on 23 June 1985. The aircraft was in transit through Irish airspace *en route* from Toronto in Canada to Bombay in India when its progress was being monitored by Shannon air-traffic control. Michael Quinn, procedural air-traffic controller on duty, had just returned from his breakfast break and had sat down to a busy 'board' or screen. Four aircraft had entered Irish airspace about the same time and had called up the Shannon tower.

Quinn assigned them codes and gave airway clearance. Three of the aircraft appeared on the radar. However, a fourth disappeared from the screen. He told his supervisor, Michael O'Hehir. It was just eight minutes after the missing plane had reported it was 'on course, all normal'. 'Air India Flight 182,' traffic controller Michael Quinn called over and over again. He radioed other aircraft and asked crews to keep a look-out for the flight. However, it had already broken into thousands of pieces in the Atlantic below when a bomb placed on board exploded inside the main cabin 31,000 feet over the south Irish coastline.

When the first rescue units reached the crash site, about 165 miles west of the Fastnet Rock, the seas were strewn with bodies, luggage, bits of aircraft and debris. Some of the passengers had apparently managed to inflate their lifejackets before finally losing their lives. The Marine Rescue Co-ordination Centre, then at Shannon Airport, had informed British and Irish military, including the Naval base at Haulbowline in Cork. Sergeant John McDermott was on an Air Corps Beechcraft surveillance plane which had just taken off on a fishery patrol when its captain received a message that 'something was in the water off Cork' and an emergency position-indicating radio beacon (EPIRB) was transmitting a navigational fix. 'We were 15 minutes in the air *en route* to the south-west fishery zone, and within half an hour we saw the wreckage in the water,' he says. The Beechcraft flew for nine hours that day, providing top cover for all the other agencies, and it also served as communications control until the RAF and the United States Air Force took over.

Alan 'Rocky' Boulden, then a flight lieutenant with the RAF at Odiham in Hampshire, was on duty that Sunday when the alert was raised. 'The search and rescue crews in Sea Kings were on 15 minutes' readiness to respond to an emergency, and I was with the Chinooks which were on a two-hour readiness for natural disasters, floods — that sort of thing. The Chinook is a heavy lift helicopter, primarily used for army support, but it is called in when a major incident occurs, such as the Lockerbie crash in December 1988 for instance. It has enough fuel to stay airborne for eight to nine hours and its range extends to 500 miles — which is much more than the Sea King.'

Boulden's was the only Chinook on two-hour standby, but the RAF managed to muster a total of three, along with their crews — on a Sunday. 'The other two followed on behind me a few hours later,' Boulden recalls. 'We ended up picking up torsos — it was mass carnage and people were just in pieces. We recovered about a dozen bodies and took them to Cork Airport in one run. There was really nothing much more we could do than that, because there was nobody to be saved.'

Working with Boulden on the Chinook ZA714 were co-pilot Flight Lieutenant Steve Ingham, winch operator Denis Gaunt and winchman Gerry Maher. 'Maher had not been trained for search and rescue, and yet there he was down among the debris and the carnage and quite a few sharks. It really was quite hazardous, and I nominated him for an award afterwards.' The Chinook flew seven hours and 30 minutes that day from its base to the search area. It flew a further seven hours and 20 minutes the following day when it returned to the crash site, but found nothing of any significance.

The sharks proved to be a bit of a nightmare also for Petty Officer Muiris 'Mossy' Mahon, Leading Seaman John McGrath and Able Seaman Terry Brown, crew of a Naval Service Gemini inflatable which was launched from the patrol ship, LE *Aisling,* under the command of Lieutenant Commander Jim Robinson. The *Aisling* had detained a British-registered Spanish fishing vessel a short time before the crash, and it was 46 miles from the flight's last reported position when Robinson intercepted the emergency message being relayed to Valentia coast radio station. Also assisting was the patrol ship, LE *Emer*, and one of two American C-130 aircraft took over the co-ordination of a seven-mile search area.

Up to 20 aircraft in all were involved, with the RAF making the biggest commitment with Chinooks, Sea Kings and Nimrod reconnaissance planes. The US Air Force Hercules C-130 transport plane supported, co-ordinated and provided refuelling for USA HH53s or 'Jolly Green Giant' helicopters, and the Royal Navy sent several Sea King helicopters from Culdrose in Cornwall. The Air Corps alternated its Beechcraft reconnaissance planes. The Royal Navy ship, HMS *Challenger*, which was designed specifically for submarine work, was on sea trials in the area by coincidence, and it also took part.

Among the many smaller craft working in tandem with the various helicopters and surveillance aircraft was the RNLI Valentia lifeboat, under the command of coxswain Seanie Murphy. The lifeboat was assigned to work with one RAF helicopter, which would drop smoke signals every so often to guide it to bodies. Crewman Nealie Lyne told reporters he had seen a solitary photograph floating in the water, and a pair of socks with the manufacturer's label still attached. A Sea King helicopter pilot, Lieutenant Gordon Jones, described the distress of seeing children's bodies in the water. 'Some of the bodies of the children were remarkably intact, with not a lot of signs of burning,' he said. 'Most of the bodies were almost naked —their clothes were shredded. Some bodies had particularly nasty lacerations and the doctor said it was consistent with them impacting the water at speed. It was consistent with a catastrophic failure at height.'[3]

Helicopter winchmen were constantly in and out of the water, while winch operators also dispersed green and yellow dye to mark specific locations. Many of the bodies were transferred to the *Aisling*. Medical orderly Jim Sperin remembers that as he approached the crash site, he was 'angry, frightened, powerless'. He felt that if someone was brought aboard alive, he wouldn't have the medical training to save them. 'We arrived at the crash site at 12.17 p.m.' Sperin wrote in an article for *The Irish Times* marking the 15th anniversary of the Air India crash in 2000. 'By 12.30 p.m. the first bodies had been taken aboard. Standing on the starboard bridge wing, I looked below to see a large woman floating by, naked, face down, her long black hair trailed behind her. The sea was littered with wreckage — mechanical and human — as far as the eye could see, and great hunks of smashed and twisted metal bobbed on the water.'

First the engineer's office on board was filled with bodies, then the carpenter's store, 'end to end', throughout the day and into the night, Sperin recalled. 'We had no body bags, so we wrapped them in sheets. Nor did we have gloves. I still recall the feeling of dead, wet flesh against mine, the stink of aviation fuel, a heavy, sickly-sweet smell that clings to flesh and metal. When the Gemini had been emptied of its cargo, what remained was hosed away. Then it was lowered into the sea and it would start all over again. Fourteen times that day the Gemini was launched; 14 times it returned, a stained and bloody white sheet covering the mound of corpses. By 10.30 p.m. there was no light left in the sky and full of bodies, 38 in all, we turned for home.'[4]

The ship spent the rest of that week as part of an extensive search operation gathering remaining pieces of wreckage while concentrated efforts were made to recover the aircraft's cockpit recorder and flight data recorder. Jim Robinson and the three Gemini crew, Mossy Mahon, John McGrath and Terry Brown, received Distinguished Service Medals afterwards.

Alan Boulden, now an RAF squadron leader, remembers that the aircraft had to be taken apart afterwards. 'The Sea Kings have a plastic floor to stop salt water affecting the superstructure, but the Chinook is not fitted with one. There were so many bodies that fluids and salt water seeped into the airframe of some of the aircraft. Floors had to be unriveted and taken out, and there was a major cleaning job. It took many, many days and many man-hours after the last search was stood down.'

The Indian families of those who perished first heard of the crash on news broadcasts. There were 84 women, 33 children and only 15 men on board. Many years later, memories of the minutes, hours and weeks after the terrible event remain with the Indo-Canadians whose first visit to Ireland was to identify the remains of loved ones — last seen in the departure terminal in Toronto.

Some still return regularly to the memorial sundial which was erected after the crash at Ahakista in west Cork. Ahakista has become as sacred to them as Amritsar is to others back in India — a tranquil place overlooking the Atlantic, linking the bereaved relatives to a local west Cork community in an effort to heal deep and searing wounds.

5

BEYOND THE LIMIT

She was like a small doll, and she would always be remembered as 'Thumbelina'. In fact Captain Dave Sparrow thought there had been some mix-up when he peered into the incubator. He had just landed the Alouette III at Tralee General Hospital in County Kerry and the medical staff were gingerly lifting the equipment into the helicopter.

'I couldn't see anything, and thought that perhaps the air ambulance had been cancelled,' Sparrow says. 'Then I saw these paper towels, and a tiny little baby in between them. They looked like king-size quilts on her, she was so small.'

It was 20 May 1986 and Eileen and Tom Lyons were distraught. Eileen had given birth prematurely at 25 weeks; she had already lost four pregnancies. The little baby weighed one and a half pounds and she was just 11 inches long, but the medical team at Kerry felt she had a fighting chance.

Sparrow remembers that the incubator had to be taken on board on a stretcher and secured. 'We were halfway to Dublin when she developed a complication. There was so much vibration in the aircraft that the medical team were terrified to touch her, so we put down in a field somewhere in the Slieve Bloom Mountains in Laois. I got into a bit of trouble over that afterwards as no one knew where we were, but we really had no choice. Once the procedure had been carried out, we took off again.'

The helicopter landed in the Phoenix Park, Dublin, and the baby was transferred by ambulance to the Rotunda Maternity Hospital. Sparrow didn't hear anything more immediately afterwards; he was about to get married himself that July. However, he received a pleasant surprise in his Christmas post some months later. The postmark was Tralee, Co. Kerry, and the Mass card stamped at the friary in Killarney was signed by Eileen, Tom and Tomasina. 'To Captain Sparrow and your team', it read. 'Many thanks for helping to save our daughter Tomasina, 1.5 lbs. You took her to the Rotunda.

She is at home now and 11 lbs. What a wonderful Xmas present!' Sparrow couldn't quite get over the child's name, though he knew she would have been called after her dad. 'You couldn't get much closer to Thumbelina than that.'

The happy story didn't end there. Some years later RTÉ Radio produced an edition of the Gay Byrne morning show out at Baldonnel as a tribute to the work of the Air Corps. Sparrow was asked to participate and was introduced to Mrs Lyons and her healthy-looking, school-going young daughter Tomasina.

1986 WAS A YEAR Captain Sparrow remembers for another reason. The Air Corps was approaching its 1,000th search and rescue call-out, having also performed hundreds of air ambulances and other mercy missions since the acquisition of the first Alouettes in late 1963. Six years before, the 1,000th air-ambulance patient had been flown from Ballinasloe, Co. Galway, to the National Medical Rehabilitation Centre in Dun Laoghaire on 9 December 1980.

'There was fierce competition between my close friend, Captain Jim Kirwan, and myself to get that 1,000th search and rescue tasking,' Sparrow says. 'We were adrenalin junkies, and now we were offering to do extra weekend duties and the like, just to try and get that magical number. I remember I did three "bucket and spade" jobs, which is the term we use for routine taskings which are close to the shore. They were numbers 997, 998 and 999, and the tension was building between us. In fact it was getting out of hand. Jim was on duty one day when the request for the 1,000th came through, and he took off. But he was stood down for whatever reason some five minutes later, which often happens if another unit has managed to effect the rescue or the alert is over. At this stage the Army press office became involved because it wanted to get a bit of mileage out of it. At the same time it didn't want any ordinary old mission.'

On 26 February 1987 newspapers carried photographs of a smiling Air Corps crew beside an Alouette III at Baldonnel — 'after completion of the 1,000th search and rescue mission', the caption read. This involved lifting 'an injured seaman from a Spanish trawler which sent a distress call 70 miles west of County Kerry in the Atlantic Ocean', it continued. The crewman was airlifted off and flown to Bantry Hospital in west Cork. The photograph showed 'the crew being congratulated by Lieutenant Colonel J. P. Kelly, officer commanding number 3 Support Wing', namely pilot Captain Dave Sparrow, winch operator Sergeant Dick Lynch and winchman Paddy

McGurk. The caption also noted that the Alouette A196 was the same aircraft that was delivered by Kelly to Baldonnel back in November 1963.

It was a significant piece of history, but it wasn't accurate! 'In fact that was the 1,006th search and rescue mission, simply because the press office needed to have a story to tell,' Sparrow says. 'Jim Kirwan never quite forgave me for that. When each of my three kids was born, he bought them a T-shirt with the same slogan, "My dad did the 1,006th!"'

PILOTS BREAK THEM; TECHNICIANS make them. The mantra among flying crews couldn't be more applicable to those involved in helicopter rescue. We couldn't do what we do without the support of the technical and engineering crews on the ground, yet they rarely get any public credit, one pilot told me, and all his colleagues agree. That support is crucial, given that helicopters 'don't like flying', as Lieutenant Colonel Harvey O'Keeffe and Captain Kevin Daunt pointed out.

'Helicopters are not aeroplanes', they wrote in an Air Corps publication to mark 30 years of helicopter operations. 'An aeroplane, by design, wants to fly. It is at its happiest high above the ground and will remain flying unless controlled by a deliberately incompetent pilot.' By contrast, helicopters are 'happiest sitting motionless on the ground', they explained. 'If they do get airborne, they are kept there by forces and controls working against one another. If this delicate equilibrium is upset in any way, the helicopter stops flying immediately. Helicopters, let's face it, are not graceful gliders. When the engine stops turning, the only way is Down [sic], straight down.'

The additional demands of the 'rotary environment' are 'quickly realised' by student pilots training in the Air Corps Gazelle helicopters, they wrote. 'From the relative comfort of hard runways and all the services airfields provide, they face tiny forest clearings for landings, navigation exercises in the worst of weather conditions, and an aircraft that wants to fly off in all directions . . .' After an initial familiarisation, training progresses to the 'dreaded hover', when students do well to 'keep the aircraft inside the dimensions of a football field at first attempts.'

This reality underlines the essential difference between fixed-wing and helicopter pilots, they noted, and it explains why aeroplane pilots are a 'bright-eyed, happy-go-lucky lot'. Their colleagues, the helicopter pilots, tend to be 'brooding, introspective prophets of doom', for they know that 'if something bad hasn't happened yet . . . it soon will!'[1]

It was both a tribute to the quality of the French-built aircraft, and the skill and commitment of the engineering crews servicing them, that the Air Corps fleet of Alouettes held such an impressive safety record over four decades. There were only a handful of 'incidents' during this period. In 1976 Alouette A195 recorded collective pitch failure over a runway at Baldonnel during a test flight and lost height instantly. In 1983 Alouette A214 *en route* to Sligo experienced tail rotor control failure and the pilot managed to make an emergency landing in a field. Both crews involved walked away unscathed. In 1980 the risk posed by low flying missions was highlighted in two incidents that year: during a joint mission with the Gárda Siochána in County Mayo, rotor blades struck overhead wires; and there was a similar occurrence in poor visibility during a search in County Louth.[2]

The Air Corps fixed-wing Beechcraft also worked silently alongside search and rescue teams and played more than a support role during an extended emergency in May 1983. The 56 foot fishing vessel, the *Árd Carna*, from Greencastle, Co. Donegal, was reported missing on 29 April 1983 when it failed to return from a trip off the north coast. The last sighting had been at 7.40 a.m. on that Friday, some 50 miles north of Inishowen Head. Its skipper reported on radio that it was about to shoot nets.

On board were skipper John Kearney, Eddie Ivers, Hugh Doherty, Martin Rudden, and Hubert Doherty — the only bachelor among them. Some 32 fishing vessels from Greencastle were joined by the Air Corps, an RAF Nimrod from Prestwick in Scotland, an RAF Sea King, the Arranmore lifeboat from County Donegal, the Portrush lifeboat from County Antrim and the Naval Service patrol ship, LE *Aoife*. The search was co-ordinated by Clyde Coastguard in Scotland.

The pilot of the Air Corps Beechcraft could scarcely believe 'the voices from nowhere' when he picked them up on radio. '*Árd Carna* drifting for four days. Unsure of position. Have not seen land for three days. No food, running low on water,' they said. There was jubilation when they were rescued. It emerged that the vessel had a leaking fuel tank and its transmitter's range was too short to send a message for help. They had had to listen helplessly as they heard the news broadcasts stating they had gone missing.

The following month, June 1983, the fishing industry monthly journal, *The Irish Skipper*, estimated that the cost of the search was

£750,000 (€952,300). A fuel leakage could be blamed on nobody, the journal noted, but many safety aids such as radio distress beacons could have alerted the emergency services to the vessel's location within hours.

The State's continued reliance on British rescue services was graphically illustrated in January 1984, when a French fishing vessel, the *Anne Sophie*, put out from Dunmore East, Co. Waterford, in a storm forecast, lost power and found itself drifting on to rocks near Mine Head. An RAF Sea King crew were faced with 90 mile an hour gusts as they winched off the eight crew in just 15 minutes and flew them to Cork. 'The seas all around were whipped up white,' Flight Lieutenant Ian MacFarlane said afterwards. 'Luckily the ship's lights were still working. It will go down as one of the more memorable jobs.'

RAF photographs of the *Anne Sophie*, its deck awash with boiling seas, appeared in newspapers the following day. Irish fishing industry representatives praised the courageous RAF crew who were put to unnecessary risk, and criticised the French vessel for putting to sea in spite of the weather warning.

GIVE WING PILOTS THE choice on a cliff or mountain rescue, and most will say they would choose the Alouette over the Dauphin — the new helicopter ordered from the same French manufacturer, Aerospatiale, in December 1982. 'The Alouette was a real workhorse, a fantastic aircraft, far superior to the RAF Whirlwinds at the time,' Captain Chris Carey says. The reliability of that single engine was never taken for granted, according to Sergeant Dick Murray. 'If it coughed, you were gone; but it never did. It was an incredible machine, the best that could be bought.'

The delay in delivering the Dauphins became political, and the then Taoiseach, Dr Garret FitzGerald, had to intervene. An Air Corps scrap-book includes a telegram from France sent by Lieutenant Colonel J. P. Kelly, dated 23 June 1986: The first two Air Corps Dauphin helicopters N244 and 246 will arrive in Casement Aerodrome, Baldonnel, on 25 June, DV.' Five years after the contract was signed, the first of the five 365 Dauphins became operational. As Dick Murray notes, it was akin to 'moving from a Model T Ford to a Mercedes' in terms of flight equipment. And these helicopters would have night-flying capability.

Two of the five were to work with the Naval Service patrol ship LE *Eithne* which had been fitted at considerable extra expense with a helipad. The other three were to be used primarily for search and rescue and other operations; in fact, like the Alouettes, the Dauphins were employed on a variety of missions including border security. And such was the twin-engine aircraft's speed, at 135 knots, that it became very popular for ministerial air transport and VIP missions, a source of some controversy as time went on.

The state-of-the-art flight control and navigation systems on the new aircraft were highlighted in the publicity surrounding their delivery; already the Alouettes were being described as 'ageing' in press reports. There was less focus on the fact that, while the Dauphins had twin engines, their search and rescue range and capacity was limited to 110 nautical miles at sea for a maximum of three survivors. This was to become apparent within four years when the first Dauphin was redeployed on a permanent basis to the Atlantic seaboard.

Within those four years several key events would determine Irish marine search and rescue policy for years to come. The first was the fate of the British-owned and Hong Kong-registered *Kowloon Bridge*, which was *en route* from Quebec to the River Clyde terminal of Hunterston in November 1986 with a cargo of 160,000 tons of iron ore, when it was forced to seek shelter in Bantry Bay in west Cork. The vessel had developed cracks in its hull during heavy Atlantic weather.

Four teams of surveyors examined the ship on 20 November while it lay at anchor off Whiddy Island. Lloyd's Register's team reported that the ship had sustained 'routine heavy weather damage' during a difficult voyage across the Atlantic. When the ship lost its starboard anchor on 22 November, it decided to set sail out of Bantry Bay; however, it lost its steerage and began to drift in heavy seas. Late on 22 November the 28-strong crew decided to abandon ship in gale force nine winds, snow and rain. It was in the early hours of 23 November in 30 foot seas that two RAF Sea King helicopters airlifted off all 28 about 15 miles off Mizen Head and took them to hospital in Cork.

The saga did not end there. Tugs couldn't reach the vessel because of the weather. After drifting for over 24 hours the ship was driven aground on the Stags rocks off Toe Head, near Baltimore in County Cork, on 24 November 1986. With its iron ore cargo, 1,700 tonnes of heavy fuel oil and 300 tonnes of diesel, it posed a serious pollution problem as it was being pounded by south-westerly seas and swells. Two of the world's top salvage companies, Smit and

Wijsmuller, couldn't refloat the ship and it broke its back on the rocks. Finally it sank on 4 December.

In the immediate aftermath it was obvious that the State had not been prepared for such an accident. Indeed most of the relevant departments were not contactable at the height of the crisis over that weekend. Had there been a coastguard and appropriate helicopters, preventative action might have been taken before the ship ran aground, it was suggested. However, not only was there no coastguard, but there wasn't even a dedicated department. Within months the then Taoiseach Charles Haughey was backing plans to establish the State's first dedicated Department of the Marine.

THAT NEW DEPARTMENT WAS still finding its feet when two Donegal fishing vessels, the *Seán Pól* and the *Paraclete*, set off on a pair trawling trip some 60 miles north of the Butt of Lewis, western Scotland, in January 1988. They had left Killybegs in south Donegal several days before on one of their first trips to sea after Christmas, and hoped to make a good run of it. Donal O'Donnell, skipper of the 120 foot *Seán Pól*, and his crew had just hauled nets, and the *Paraclete* was steaming away to pick up another mark when trouble struck. 'We had a bag of fish, and crewman John Johnson was pulling on the lifeline. He took a turn on the rail and put his hand in the loop. Suddenly the boat took a roll, broke his arm which was caught in the loop, and pulled him overboard,' O'Donnell describes.

It was dark, there was a rolling swell, and O'Donnell knew that 12 minutes was about as much as anyone could endure in the water at that time of year. 'I put a man on top of the wheelhouse and I put the searchlight on him in the water, but the bulb fused. We lit flares so we could keep sight of him, and there was a smoke flare on the lifebelt we threw him. However, it was blowing into his mouth and choking him, and he was trying to shake it off. Luckily he didn't succeed.'

The *Seán Pól* had used up its last flare when the *Paraclete*, skippered by Dinny Byrne, caught sight of Johnson in the water. During that time Johnson had stayed very calm and only panicked when he saw the *Paraclete* apparently heading straight for him. 'Dinnie Byrne tried to reach John and they had a grip on him twice. They would have him up to the level of the rail on the deck, and then fierce suction would pull him down again. One of the *Paraclete* crew grabbed the grappling hook, which was used for pair trawling, and somehow it caught in Johnson's trousers and held him. I was just saying a prayer to St Catherine when the *Paraclete* confirmed that it had him,' O'Donnell says, with a clear memory of the absolute relief he felt at that moment.

When Johnson was pulled aboard it was obvious he was in severe pain and approaching hypothermia. He had been 20 minutes in the icy seas and had sustained severe arm injuries. He would need urgent medical help. The vessel put out an alert, and Stornoway Coastguard scrambled a British Coastguard helicopter which was based on the Hebrides. Pilot of Rescue 119, Captain John Bleaden, was airborne shortly after 7 p.m. It was a horrible night, the forecast was not good, but he and his three crew had over 70 years of flying experience between them.

In fact storm force winds were gusting to 70 knots when the helicopter arrived over the *Paraclete* west of the Butt of Lewis. The vessel was at least under power, and this should have made the helicopter crew's job a little easier, but initial attempts to lower winchman Jeff Todd using the hi-line technique didn't succeed. The line was dropped on board the vessel, attached to Todd, and he was lowered, with the *Paraclete* crew trying to pull him on to the deck. However, the vessel was rising and falling over 60 feet, twisting like a corkscrew in the heavy seas, and Todd was in danger of being hit. He was winched back on board, and the pilot decided to try a straight winch to the deck as the only option.

'One minute I looked around and the deck was 30 feet below me, the next I was looking up at the screws of the trawler,' Todd said afterwards. 'I hit the deck and was dragged across it and down into a well. I was now under the deck and if I hadn't managed to disconnect on time, I would have been dead. Once I got down on board the ship and went to talk to the skipper, I couldn't. My tongue was stuck to the roof of my mouth with fear.' He described it as the worst experience in his 28 years of flying. However, he had to pull himself together and tend to the injured crewman, while the helicopter moved away. He made Johnson an improvised sling for his arm and strapped him tightly into a stretcher. Once he was ready, Todd gave the *Paraclete* crew brief instructions on how to work the hi-line.

'There was no way we could have winched them straight off the deck,' Bleaden recalled. 'We had to stand back or else they would have gone into the superstructure.' As it was, the winching operation proved to be a terrifying ordeal. The pair were swinging so violently on the cable as they were being hauled back up that at one point Bleaden and his co-pilot Andy Hudson saw Jeff Todd and the casualty right in front of them. 'We watched Jeff go by in a line with the tail rotor over the helicopter nose. I've never seen that before,' Bleaden said.

'When we came off the deck, we were swinging so violently that the wire caught the undercarriage of the chopper and the hi-line became looped around the casualty's neck in the stretcher,' Todd said. 'We were moving about so much that it would have throttled him, so I cut the line with my knife and then continued safely to the chopper.' The winch operator, Vic Carcass, said he had only seen conditions similar to that night once before in his flying career, when the RNLI's lifeboat at Fraserburgh had capsized and he was among the helicopter crew called out.

About 30 minutes after leaving the aircraft, Todd was back on board with Johnson still strapped to the stretcher. Bleaden flew directly to the hospital in Lewis; with the storm behind him, the journey was completed in half the time it had taken to get out there in the first place. Johnson was treated and flown later by air ambulance to Glasgow.

'That rescue was right on the limit,' said Bleaden. 'I'd never have attempted it with a less experienced crew, and even then maybe we shouldn't have. But once you've made up your mind, you just have to go for it. The crew were great. We could never have done it unless we'd worked as a team. My co-pilot Andy Hudson had one of the worst jobs because he couldn't see a thing out of the cockpit, he was so busy giving me information on performance and conditions. I said in my report on the mission that Todd deserves a medal — and I need my head examining!'[3]

The Killybegs Fishermen's Organisation decided it should make a presentation to the British helicopter crew for a most heroic rescue. Joey Murrin, chief executive of the organisation, said he would be writing to the Secretary of State in Scotland to nominate the men for the highest gallantry award. He had also sent an account of the rescue to the government, the Taoiseach, Charles Haughey, and the Minister of State with responsibility for fisheries, Pat the Cope Gallagher, in the hope that there would be national recognition of 'this feat'.

Several months later the Stornoway helicopter was involved in a crash during a rescue off a Scottish island; the pilot, Captain Bleaden, and his crew survived, but the aircraft was scrapped.

6

THE WEST COAST CAMPAIGN

Donegal fisherman John Oglesby was a skipper who commanded enormous respect among his peers. Born on Owey Island, he had fished for most of his life and was in his early forties when he acquired the 120 foot vessel, the *Neptune*. At the time it was one of the largest and most modern vessels in the fleet.

In the early hours of 12 February 1988 the *Neptune* was fishing off the Mayo coast in good conditions when a piece of equipment failed. The skipper was on deck when a sheave pin lifted, loosening a trawl warp which severed his leg. Crew member Tom Rawdon was also injured. Malin Head coast radio was alerted; the nearest lifeboat station was Arranmore Island, Co. Donegal, and the Air Corps rescue squadron was on the east coast at Baldonnel, so a call was put through to Britain. However, by Royal Air Force calculations, the vessel would have reached port before a helicopter from Prestwick could reach it.

The *Neptune* was four hours' steaming time from Broadhaven Bay in north Mayo. Among the crew trying desperately to help John Oglesby was his son Martin. Three and a half hours later the skipper lost his battle. He was within sight of land when he died.

Joan McGinley, mother of four young boys and married to fisherman Mick McGinley, was at home in Teelin, north of Killybegs in County Donegal. She was angry. Only three weeks before, Killybegs fishermen's leader Joey Murrin had spoken publicly about the need for a rescue helicopter on the west coast, as both he and Dr Marion Broderick, general practitioner for the Aran Islands, had done for several years. Murrin's radio interview concerned a statement in which he had criticised politicians for 'joy riding' around the country in Air Corps helicopters, while those who toiled off the Atlantic seaboard were dependent on Britain for air-sea rescue.

As McGinley recalled, the RTÉ Radio interviewer Gay Byrne had been initially sympathetic, but several days later Mr Byrne returned to the subject and was far less so. A female listener had phoned the radio programme in the interim to point out that there was a fine helicopter already based up in the north-west coast at Finner Army camp. Byrne reacted as if he felt he had been taken in.

Joan McGinley wrote to the radio programme to defend Murrin. The helicopter at Finner was a single-engine Alouette III which could not work over water without top cover and in limited range, she pointed out. She also had a tea towel hanging over her washing machine which clearly showed the concentration of lifeboats on the east and south coasts. There were only three stations along the whole of the western seaboard.

Now, just several weeks after that exchange, a leading fisherman had died unnecessarily. McGinley got working. She organised a public meeting for 31 March in Killybegs. Fishing industry representatives from Dingle in County Kerry to Greencastle in north Donegal pledged support. Posters showing 23 lifeboat stations on the east coast and three on the west, with five Dauphin helicopters based at the Air Corps base in Baldonnel were distributed, and 33 elected politicians were invited to attend. 'It's so rarely that the fishing industry unites on anything, but it will on this one,' Mrs McGinley predicted. She was right.

The turnout for the meeting was overwhelming, and the consensus from the floor was that Joan McGinley should form a committee of her choosing. Several people had contacted her beforehand, including some individuals like Dr Marion Broderick of the Aran Islands who had raised this issue of inadequate cover on the west coast before. Being a trained community worker, McGinley made mental notes of the various skills she would require on her committee to ensure that it was effective. 'I knew I needed helicopter expertise, a master mariner, someone from the Naval Service, someone representing fishing interests, medical expertise and legal input. It had to be small and geographically spread,' she says.

And so a committee was formed, representing a wealth of experience. Included were Commandant Fergus O'Connor, formerly of the Air Corps and with Irish Helicopters; two ex-commanders of the Naval Service, Eamonn Doyle and Paddy Kavanagh; Joey Murrin, chief executive of the Killybegs Fishermen's Organisation; Bryan Casburn, manager of the Galway and Aran Fishermen's Co-op; Dr Marion Broderick, GP Aran Islands; Peter Murphy, a solicitor from Letterkenny; and Joan McGinley in the chair.

The budget was just over £120,000 (€152,000) which was made up of individual and corporate donations. The people of Inis Meáin held a door-to-door collection raising £337.09 (over €400) which sustained the committee's work when funds got tight. Within a few short months the voluntary committee had published a comprehensive report on the current state of search and rescue off the Irish coastline, largely written by Eamonn Doyle and Fergus O'Connor. It highlighted the urgent need for action. As a coastal state, Ireland was obliged to establish, operate and maintain an effective coast watch and search and rescue service under international law. The frequency with which British helicopters responded to taskings in the designated Irish search and rescue zone — which was equivalent to the air-traffic control zone — showed that this international obligation was not being met.

In 1987 for instance, the Irish Marine Rescue Co-ordination Centre (MRCC) in Shannon drew on British air-sea rescue services on 54 occasions, compared to 80 call-outs for the Air Corps. The report identified flaws in the current organisation: the location of most resources on the east coast; the limited operational range of the Air Corps Dauphin helicopters; the lack of modern radio communications equipment in MRCC — without even a VHF or medium-frequency radio and only telex contact with the coast radio stations at Malin in Donegal and Valentia in Kerry; the lack of procedures to review national search and rescue plans; and the lack of a coast watch, in spite of repeated calls by maritime historian and Dun Laoghaire lifeboat honorary secretary, Dr John de Courcy Ireland.

The report noted the 'extraordinary' situation where the Naval Service flagship, LE *Eithne*, sailed on patrol without its Dauphin helicopter — four years after commissioning. It had been constructed with a helipad on deck to work with the Dauphins. This greatly diminished search and rescue capability, it said. It noted that the Air Corps Dauphin was a short-range rescue aircraft, and there was a large area beyond its radius of action where effective search and rescue cover could only be provided by the Sea King, an aircraft which could also carry up to 20 survivors, depending on range. As there appeared to be no plans to purchase long-range aircraft for the Air Corps, the British model of inter-mixing public and private services to provide search and rescue (SAR) should be looked at. 'Even though the British are extremely well equipped with SAR designed military helicopters, they have felt it necessary to station commercially operated SAR helicopters at three locations around their coast at Stornoway in the Outer Hebrides, Sumburgh in the Shetlands and most recently at Lee-on-Solent in the English Channel because they considered that the civil SAR need would not be otherwise met.'[1]

The report recommended that the Air Corps search and rescue squadron should be based at Shannon, with one helicopter detached on a rotational basis to Finner, Co. Donegal; that these Dauphins be ready to fly in 15 minutes by day and in 45 minutes by night; and that a long-range search and rescue helicopter should be stationed on the west coast to allow unrestricted operation out to 15 degrees west longitude and with a capability for up to 20 survivors. While the Irish search and rescue region on the east, south and north-west coasts was within 50 nautical miles, the west coast boundary, set at 15 degrees west, was 220 nautical miles from the nearest air base and way beyond the operating radius of any helicopter then used by the Air Corps.

The report also recommended that a coast watch service should be established; that emergency position-indicating radio beacons (EPIRBs) should be fitted on all commercial seagoing vessels and all pleasure craft above a certain size to give a satellite fix in case of emergency; and that a new rescue co-ordinating structure be established with proper communications at the MRCC in Shannon and clearly defined sectors of responsibility between Shannon, Malin and Valentia. And it said that emergency air-ambulance missions from the 21 inhabited islands off the west coast of Ireland should be considered as 'search and rescue' and given the same priority.

The report was presented to the Minister for the Marine who had set up an interdepartmental committee on deploying Dauphin search and rescue helicopters several months before. It could not agree on firm recommendations, having studied the McGinley committee report. So in February 1989 the minister set up his own review group, chaired by retired Garda commissioner Eamonn Doherty, and with a line-up that included none of the McGinley committee members, but many who were experts in their respective fields. International representatives included a European Commission official, a US Coastguard commander and a former regional controller with the British Coastguard.

In the intervening weeks another spate of incidents highlighted the pressure on existing services. On 13 January 1989 three fishermen were drowned when a Spanish fishing vessel, the *Big Cat*, ran aground in heavy seas off Valentia. Winds were gusting to gale force nine when the distress signal was picked up by Valentia coast radio at 7.34 a.m. The vessel had been trying to approach Valentia Harbour when it was driven towards Beginish Island and thrown broadside on the rocky shoreline. The captain and engineer were among the unfortunate three who made for the shore when the seas shattered the vessel's wheelhouse windows.

Conditions were so bad that the Valentia lifeboat could not make a safe approach, and the *Big Cat* was taking in water. An RAF Sea King from Brawdy in south Wales was *en route*, but the local coast lifesaving service managed to use nineteenth-century breeches buoy equipment to winch the 11 survivors ashore. The five volunteers with Valentia Coast and Cliff Rescue Service — Michael O'Connor, Patrick Curtin, Owen Walsh, Aidan Walsh and Peadar Houlihan — were recognised for their courage subsequently by Comhairle na Míre Gaile, the awards for bravery council, at a ceremony in Cahirciveen district court on 16 July 1992.

Ironically the British Transport Secretary had decided to drop the continued use of breeches buoy equipment on the British coastline, and Ireland had purchased some of it at a bargain price. After the *Big Cat* incident, Captain Peter Brown, superintendent of the Coast Life Saving Service, said the equipment would continue to be used in Ireland until such time as there was adequate helicopter rescue cover.

Even as the *Big Cat* rescue was under way, a British cargo vessel, the *Gladonia*, had gone aground in Tramore Bay on the south-east coast, and the *Yarrawonga*, a bulk carrier with 32 crew on board, was reported to be in danger of sinking some 400 miles off the Irish coast, while a light aircraft had crashed on the Aran Islands several nights before. The West Coast Search and Rescue Action Committee criticised a decision by the MRCC to scale down the search for the aircraft before positive information on its whereabouts. The MRCC did not have direct radio contact with any of the rescue aircraft, lifeboat personnel or fishing vessels assisting in the search.

The United States dispatched several 'Jolly Green Giant' rescue helicopters to the *Yarrawonga* to winch off the crew. Once abandoned, the tanker did not sink and was at risk of drifting on to the west coast. It was around 170 miles west of the Aran Islands when the Naval Service flagship LE *Eithne*, under Commander John Kavanagh, was sent to locate it, but was unable to put a boarding party on the vessel. Kavanagh sought the assistance of the Air Corps

— after all, the *Eithne* had been built with a helicopter deck.

The Air Corps Dauphin was working at extreme range, in fresh weather, and conditions made it impossible for it to land on the *Eithne's* deck. Instead, it refuelled while airborne and hovering over the ship in a procedure known by its acronym, HIFR (helicopter in-flight refuelling). A salvage party from the *Eithne*, including Lieutenant Commander Gerry O'Donoghue and Leading Seaman Kieran Monks, was then winched on board the 'steam chicken', as the helicopter was nicknamed, and transferred on to the tanker. Two Filipino crew from a Dutch salvage tug, the *Typhoon*, were also taken on board to assess the scene. The *Typhoon* had last been in Irish waters when an abandoned ship, the *Kowloon Bridge*, hit rocks off the Irish south coast in 1986.

As O'Donoghue recalls, the ship was holed just on the waterline and at least one of its tanks was 'free flooding'. It was in a poor state of repair but not in danger of sinking. He remembers there were signs of looting in the crew quarters — suggesting that the ship's officers had lost control when the vessel lost power.

Naval divers from the *Eithne* then tried to climb up the side of the ship using some grappling lines left by another vessel, possibly a fishing crew which had been trying to get on board earlier. The long climb on small lines proved too difficult. Finally a line was secured on board the ship by the Dutch tug. The Naval boarding party was taken off by a searider inflatable, and the *Eithne* stayed put until the *Typhoon's* tow was under way at a speed of about four knots.

While the operation was a success from a Navy/Air Corps point of view, it also highlighted the limitations of the Dauphin; ironically one of the factors influencing the craft's purchase was the fact that it would be able to work with the navy ship.

The Doherty committee produced an interim report which recommended that an Air Corps Dauphin be relocated to Shannon and said that Finner camp in County Donegal should be 'increasingly used' for search and rescue operations. It also called for the re-equipping of the MRCC and the streamlining of alert procedures between the MRCC, coast radio stations and rescue agencies. Moves were then made by the government to implement these measures. The committee's final report, signed off in February 1990, recommended that a new division of the Department of the Marine should be established to assume responsibility for maritime safety, rescue, shipwreck and sea and coastal pollution, called the Irish Marine Emergency Service.[2] This would take over the activities of the MRCC, the coast radio service, marine radio engineers and coast and cliff rescue service, and would be headed by a director answerable to the Minister for the Marine. The MRCC should be moved from Shannon to Dublin and three marine rescue subcentres should be established at Dublin, Valentia and Malin Head.

The report recommended that two medium-range helicopters — Sea King or similar—should be purchased for the Air Corps, and that this service should be located at Shannon. Pending purchase, delivery and deployment of these craft, a private contract to provide medium-range helicopter cover should be negotiated 'as a matter of urgency'. Once this contract was based at Shannon, the Dauphin service provided there should be moved to Finner camp on a 24-hour response basis.

The report highlighted the contribution made by the British military helicopters in the Irish search and rescue region over the years, principally by RAF Sea Kings from the south Wales military base at Brawdy. It praised the professionalism of the RAF and noted that the military wing was eager to continue this service. However, it also noted the disadvantages of the arrangement, including a time delay of one to two hours' transit from Brawdy to Shannon, and the fact that the RAF could not provide the service where safety of life was not directly at stake — as in the case of the *Kowloon Bridge* in 1986 and the *Yarrawonga* in 1989. And there was also the fact that the RAF's first priority was military search and rescue. In cases where one or two Sea Kings at Brawdy went out of service, the remaining one craft might not be available for Irish operations. The British Coastguard had had to contract out some of its service to civilian helicopters for this reason.

There was much more in the 54 recommendations. The final report mirrored that produced by the West Coast Search and Rescue Action committee, and represented a vindication of the campaign fought by Joan McGinley and her committee. A government contract for a medium-range service at Shannon was drawn up, even as the Air Corps continued its duty there with the Dauphin; the recommended purchase of new craft for the defence wing, which the Doherty committee had advised 'as soon as possible', was put on the long finger.

The government's decision to concede Shannon as a west coast air-sea rescue helicopter base was to prove its worth just a month after the Doherty final report, when the Air Corps search and rescue crew, relocated to the west, were involved in a most dramatic mission. 'I have never seen such a sea state in my flying career [15 years],' Commandant Jim Corby noted afterwards.[3]

The distress call relayed by the MRCC came shortly after midnight on the night of 8/9 March 1990. A 65 foot fishing vessel, the *Locative*, with four crew on board had lost engine power and was taking in water somewhere off Arranmore Island in Donegal. Commandant Jurgen Whyte, Dauphin commander on duty, alerted the crew — co-pilot Commandant Corby, winch operator Sergeant Ben Heron and winchman Corporal (now Flight Sergeant) Daithi Ó Cearbhalláin, whose reputation went before him as a former soldier with the Army Ranger wing.

'Yogi', as Whyte was known to his colleagues, was Dublin born with a German mother and had joined the Air Corps in 1976. He was one of the search and rescue unit's most experienced pilots, having flown initially in fighter squadron jets, in helicopters, and having worked as an instructor. He had held several key posts, including officer commanding the Naval Service support squadron, and officer commanding search and rescue.

Earlier that evening the crew had abandoned a night winching training exercise due to bad weather. A north-westerly gale was gusting to severe gale force nine, with seas of three metres and a very heavy swell of up to 10 metres (over 30 feet) in height.

Whyte was concerned about the wind conditions, the sea state, and the fact that the MRCC was not sure of the vessel's position. He was also well aware of the limitations of a small machine working in big Atlantic seas. He asked the MRCC for an RAF Nimrod to help in the search and a Sea King from Britain as a possible back-up; the Arranmore lifeboat was also *en route*.

Corby got a detailed briefing from the Meteorological Service, and it appeared that the weather would only deteriorate north of Ballina. The worst conditions would be in and around Arranmore Island with winds of over 70 knots and a heavy rolling sea. The captain decided to fly to Finner, refuel and reassess the situation there.

The Dauphin took off from Shannon for Finner at 1.55 a.m. and by Castlebar it had made contact with the RAF Nimrod. Flares had been sighted by another fishing vessel south of Arranmore in and around Rathlin O'Beirne. However, the helicopter crew were under pressure as the MRCC had informed them that the RAF Sea King had had to turn back due to icing weather conditions. The Dauphin made visual contact with the Nimrod when it reached Sligo Bay at about 3.10 a.m; the Nimrod was checking out three possible 'targets'. Nothing had been heard from the fishing vessel for 30 minutes.

By chance the Dauphin heard the *Locative* call on VHF channel 24. Using direction-finding equipment, the Air Corps crew were able to estimate its position to the west. Several minutes later both the helicopter and the Nimrod spotted a red flare. There was no time to refuel; in any case there was enough fuel for 90 minutes' flight time. The helicopter flew out to the vessel which was drifting broadside in an enormous Atlantic swell. The crew of four were huddled at the stern of the heaving vessel, which was fortunately visible under a full moon. All were wearing lifejackets, a rare enough occurrence in many such emergencies at the time.

It had taken one hour and 15 minutes to get there, but the work was only beginning. The aircrew spent another 30 minutes trying to hold the aircraft over the vessel to allow winchman Ó Cearbhalláin down safely. As Ó Cearbhalláin recorded afterwards, the pitch and roll of the vessel was the worst he had seen to date during his career. He had to take account of the gear on deck, including a large ship's aerial, a derrick at the bow and several lines and aerials running between it and the wheelhouse.

'The vessel itself was rising and falling 80 feet in the swell,' Whyte says. 'This successive rate of change exceeded the capability of the Dauphin's automatic hover system.' The hover system allows the pilot to set the minimum height between the belly of the aircraft and the sea; the aircraft will rise and fall with the swell — and automatically fly away if that sequence is broken.

The captain decided to fly the helicopter with manual height control, which is extremely risky and increases the workload of the crew; the co-pilot has to call heights continuously to the pilot, who cannot see the swell, and the winch operator has to call out warnings as and when the helicopter comes close to the water. 'As winch op., you are the one looking out the door and you have an unrivalled bird's eye view of the whole situation,' Ben Heron says. 'You can see the clearances. It is the pilot's job to do what he or she is told, and it is all based on trust.'

There was an additional risk factor: the aircraft was at constant risk of being skewered by the mast of the vessel below. 'Due to the wind position of the *Locative*, I couldn't see it below me and under these conditions the chance of collision is very high,' Whyte says. He decided to stand off and wait for the arrival of the Arranmore lifeboat, leaving the four men being tossed about on the deck wondering what was going on.

Within 15 minutes the Arranmore lifeboat arrived — to the relief of the Dauphin crew. It was now 3.35 a.m. Over the radio the aircrew explained that they couldn't attempt a lift with the vessel lying parallel to the swell and at 'cross decks' to the helicopter in hover. The lifeboat made several unsuccessful attempts to approach the *Locative*. At one stage Whyte recorded, 'We witnessed the trawler bearing down on top of the incoming lifeboat', and only 'prompt, evasive action' by the coxswain averted a collision. 'We thought the lifeboat would be able to come alongside and drag the guys off the deck. Instead, we witnessed this incredible sight where the lifeboat was trying to dart in to the vessel and the *Locative* would rise up over the swell and fall down towards it. The coxswain was incredible, but we knew then that the lifeboat wasn't going to do it.'[4]

There was just 40 minutes of hovering time left. The fishermen were totally dependent on the helicopter, which was running short of fuel; if it flew in to refuel, the four men might not survive. The pilot and winch operator remembered reading an account of a rescue where a lifeboat had pulled a powerless vessel around. 'Picture the situation where the helicopter was hovering north-south, and the vessel was lying east-west. If the vessel could be pulled into a north-east position, we could at least see part of it, Whyte says.

The aircrew suggested that the lifeboat try and get a line aboard and pull the vessel to a 30 degree heading off wind which might be enough to provide visual clues for the helicopter in hover. Coxswain John O'Donnell managed to get two tow lines on board and manoeuvred the *Locative* successfully into position. 'Once the vessel was lying at this 30 degrees offset, I could see a pattern,' Whyte says. The two vessels — lifeboat and *Locative* — were engaged in a surreal dance across the swell, to the extent that the captain could anticipate the movement of one by the other. 'That sequence developed a distinct pattern, and this allowed us to go in safely.'

At this point the winch crew lowered the hi-line, hoping that the fishermen would know what to do with it and wouldn't secure it to anything on the deck! Ó Cearbhalláin — 'the dope on the rope' as his colleague, Heron, describes the job — then descended and within a few minutes he had sent one of the crewmen up. 'Due to the big swell the finer points of winching, for example winching in slack cable and gently winching the survivor clear of the deck, were discarded and the survivor was "snatched" off,' Ben Heron said afterwards in his report on the mission.[5]

Once Heron had hauled the first survivor into the helicopter, he winched the strop back down to Ó Cearbhalláin. The hook got caught in a fishing net, but the winchman freed it and placed the second crewman in the strop. However, 'at this point things started to go wrong,' Heron said. A large wave hit the boat, throwing it up towards the helicopter and snapping one of the two tow lines from the lifeboat. The captain had to climb rapidly and move back to avoid being hit by the ship's aerial. Heron winched out as much slack as he could to prevent the second crewman from being dragged off the deck when the boat went over the top of the wave.

With one tow line gone, the coxswain had to reduce his towing speed to maintain the second line. If it snapped, the vessel was gone, and there were still two people on board. However, in reducing the tow the lifeboat and helicopter had to cope with a more erratic and haphazard motion from the stricken vessel. Winching became all that much more difficult.

Simultaneously, Ó Cearbhalláin noticed that the 'weak link' on the hi-line had broken, and it was no longer attached to the hoist hook — a very rare occurrence. With great presence of mind, he stuffed a bundle of the hi-line into the strop with the second crewman, just as the boat slid down the back of a wave and the crewman was dragged off. He was scooped up in a massive swing and the hi-line tangled around him. Heron remembers he only knew he had him when he felt the shock coming back up through the cable. 'He spun around and got all caught up in the hi-line,' he says.

The winch operator untangled the hi-line furiously, as he had no knife to cut it. He then had to replace the 'weak link' with one from a spare hi-line, and winch the strop back down to Ó Cearbhalláin. It took a good ten minutes to make the repair. Fuel was running low and the winchman was getting anxious. The delay seemed like an eternity, according to Corby. His colleague, Whyte, had to maintain a hover which was 'too close for comfort' over the vessel, without the vital assistance of 'patter' from the winch operator. 'The strain on all concerned was particularly severe as we had been in the manual hover for over an hour in the worst conditions any of us had ever seen.' The crew didn't know for how much longer the hoist would hold out in the violent snatch lifts.

With just 25 minutes of fuel left, winching resumed, and the third 'snatch lift' was as hazardous as the previous two. Shortly after the last fisherman was taken off and the lifeboat was towing the

Locative, the second tow line snapped and the vessel was left to the mercy of the sea. The helicopter routed directly to Finner with the four fishermen and landed with just five minutes of fuel remaining. Coxswain O'Donnell later told the Air Corps board of inquiry that it was a 'hellish night', a fitting statement, the Air Corps noted, from a man who had received a citation for his courage from the RNLI.

The aircrew agreed that the lifeboat's presence was crucial in helping to position the fishing vessel and in acting as a visual reference. Both crews had demonstrated much courage, stamina and seamanship. The pilots knew that the winching crew were the very best they could have hoped for — 'top guns', Whyte remarked afterwards.

The vital need for constant radio communication between winchman and aircraft was raised by members of the aircrew in their reports to the Air Corps. For their efforts they were awarded a Distinguished Service Medal with distinction, the first time a Dauphin crew had been recommended for one. Given that only 17 DSMs have been awarded to Air Corps flying personnel since 1963, it was particularly significant. It was also the first such medal for a sea rescue, and the first night rescue by a Dauphin attached to the Air Corps fleet.

A group of 57 French fishermen and processors also owed their lives to the early establishment of the Shannon base. In the early hours of 5 April 1991, a 270 foot French factory ship, the *Capitaine Pleven II*, ran aground on the Loo rock about two to three miles east of Black Head on Galway Bay's southern flank. Commandant Harvey O'Keeffe, now promoted to Lieutenant Colonel, remembers the call-out. A training mission scheduled for that night had been cancelled due to the weather.

'We got a phone call to say a ship was aground with 57 on board.
'"Seven?", we said.

'When they confirmed the total, we did suggest that a couple more aircraft and rescue units might be useful!

'We took off from Shannon in atrocious conditions, with 60 knot winds, and it was slightly difficult getting up and in there,' O'Keeffe says. It took about 40 minutes. 'However, because the ship wasn't moving, it was textbook in a sense. There was some turbulence from the mountain which affected winching, and constant driving rain. Other than that it was pretty straightforward.'

Flying with O'Keeffe were Captain Sean Murphy, Corporal Christy Mahady and Corporal Daithi Ó Cearbhalláin. Being closer to the scene, the Dauphin had carried out the first airlift from the ship by the time the RAF Sea King arrived from Brawdy, south Wales. 'We estimated that we could take six, but when we took off and looked in the back, we had seven on board!' Murphy says. All the crew were wearing survival suits and lifejackets.

Flight Lieutenant Mike Baulding was captain of the Sea King, along with Pilot Officer, Flight Lieutenant Richard Hooper, Warrant Officer and winch operator Peter Williams, Flight Sergeant/winchman Mark Stephens and aircraft technician Ian Slater. The two helicopters took turns ferrying crew members to Carnmore Airport in Galway city 30 miles away, as the ship lay broadside in turbulent seas off the Loo rock with a 25 degree list to seaward and three holes below its waterline. Even as they continued the rescue, the helicopter crews noticed a small oil slick spreading from the vessel.

The first to be taken ashore were the 47 workers engaged in processing the fish catch below decks; meanwhile Captain Jean Marc Le Borgne and his fishing crew made several efforts to refloat the vessel. Fire brigade units from Clare and Galway had arrived in Ballyvaughan, Co. Clare, to supply pumping equipment which might refloat the ship. However, before the pumps could be put into action, the ship's list increased. The captain and crew decided to abandon any further attempts, and all were airlifted safely ashore.

The airlifts took six hours and more. 'We did three runs; the RAF did two,' says Murphy. 'When the emergency first arose, the Whitethorn Restaurant in Ballyvaughan switched on all its lights to act as a beacon for the ship. It was a throwback to the time when the cut-stone building had been a British coastguard station, Michael Finlan of *The Irish Times* wrote.[6] The pilots — Harvey O'Keeffe of the Air Corps and Mike Baulding of the RAF — played down their efforts when interviewed afterwards by Jim Fahy, RTÉ television's western correspondent.

Warrant Officer Peter Williams from the RAF Brawdy crew agreed that it had been very smooth. 'It's always a great pleasure for us to come to Ireland,' he said. The bar was opened for the ship's crew at Carnmore Airport. One of them, Cedric Dore from Brest in France, said there was 'no panic', and everyone 'kept cool'. 'We're very happy at the reception we got here.'

Daithi Ó Cearbhalláin, winchman on that night, remembers that the mood was calm on the deck of the ship and he even had time to have a few photographs taken with his counterpart from the RAF, Sergeant Mark Stephens. Having close contact with the crew on the winch, he also recalls a distinct smell of wine among some of them! 'Afterwards one of the crew wanted to give me a big smacker. I just wasn't in the mood!'

Once all the crew were safe and accounted for, the ship became a pollution threat. A pollution-control operation for Ballyvaughan Bay and the south Galway Bay area was mobilised; the ship had an estimated 600 tons of gas oil and 22 tons of lubricating oil on board. There were genuine fears that it might leave an indelible mark on the north Clare coastline — the sea border of the dramatic limestone landscape of the Burren. Symbolised by the gentian, the 250 square kilometre limestone pavement formed hundreds of millions of years ago runs down on to cliffs and terraces embracing a rich and diverse Atlantic marine life, extending from Black Head to Doolin. North Clare also had a developing shellfish farming industry. One could only imagine then, the extent of concern about the impact of any fuel spillage on the fragile seaboard.

There was also speculation as to how the grounding had occurred, in spite of the bad weather. The St Malo-owned stern trawling factory ship was built to withstand the worst of Atlantic weather and would have been equipped with radar, sounder, plotter, weatherfax and other electronic navigation aids. Although Admiralty charts for the west coast were outdated, the rocks on which the vessel had grounded were clearly marked.

The ship's agent said that the *Capitaine Pleven II* had been fishing for the previous month on the Porcupine Bank and had put into Galway Bay to leave an injured crewman ashore. When weather conditions deteriorated *en route*, a rendezvous for this purpose was postponed and the vessel headed for an anchorage off Black Head. Local fishermen pointed out that the anchorage was only safe in prevailing south to south-westerly winds. A salvage contract was organised, but the drama took another twist when the tug involved was detained by customs officers at Galway docks. Eventually the vessel was taken into dry dock at Verolme shipyard in Cork for structural repairs. Curiously, the then marine minister, John Wilson, did not order an inquiry as he regarded the grounding as being 'purely accidental'.

The Air Corps and RAF crews were decorated by both Britain and France for the rescue. They received the Edward and Maisie Lewis Award from the Shipwrecked Fishermen and Mariners' Benevolent Society in 1992, and in the same year they were presented with a certificate from Jean Yves Le Drian, French Secretary of State for the

Sea. In his citation, Monsieur Le Drian conveyed *felicitations* and *remerciements* for the courage and professionalism shown by the men in saving the crew of the ship during a 'violent tempest' in Galway Bay in 1991.

7

CLOSE CALLS

A tranquil Dublin Bay was bathed in the silken light of a full moon on 21 November 1991, one of those rare, magical winter nights. It was an hour before high water on a spring tide and the crew of 16 on the MV *Kilkenny*, a B & I Line cargo ship, were looking forward to berthing on the Liffey after a voyage from the Belgian port of Antwerp. If there were no delays they might just catch a pint before the last bus home.

Leaving the Liffey mouth was a German-registered container ship, the *Hasselwerder*, which had a crew of 14 on board and, by pure coincidence, was bound for Antwerp. At about 9.45 p.m. both ships were about a mile and a quarter east of the Poolbeg lighthouse, between number three and four buoys in the designated shipping fairway. The second mate of the *Kilkenny*, Anthony Flanagan, was off watch and below decks reading when he heard a loud crash and the vessel began listing heavily to port.

Flanagan rushed to the bridge, grabbed his lifejacket and then ran to the deck. Feverish attempts were being made to launch the ship's liferafts even as containers on the ship began sliding into the sea. However, the liferafts on the port side were already under water and the rafts on the starboard side couldn't be released because of the severe list, which Flanagan estimated was about 80 degrees at this stage.

He later described how he and four other crew jumped from the deck, and more of his colleagues followed. 'Everything was "black out", and the water was very cold. Some of the men couldn't swim, but everyone obeyed instructions to keep in touch by shouting,' he said. While trying to keep afloat, and together, some of the crew roared to the *Hasselwerder*, which was within 100 yards of the stricken ship and with a bow clearly damaged by the impact. They claimed there was no response to their pleas, but it emerged later that the *Kilkenny*'s captain was picked up by a liferaft from the *Hasselwerder*.

One of the men in the water, Austin Gill, had no lifejacket. Keith

Marry, 19-year-old deckhand on the ship and a former fisherman, called him over and shared his jacket with him. They clung together for about 15 minutes until the ship's chief engineer swam over and the three used two lifejackets for buoyancy. Marry said that the possibility that he might not survive did go through his head. 'But you can't think like that. You just have to hope.'

They were in the water for about 20 minutes when they were picked up by the RNLI lifeboat which had launched from Dun Laoghaire, and by the MV *Leinster* passenger ferry. As the *Leinster* approached, Marry remembers that one of the men in the water started singing 'Here we go, here we go, here we go', and the rest of them joined in.

In fact some of the men may have been in the water longer, because questions were asked afterwards about the delay in calling out the lifeboats from Dun Laoghaire and Howth. The secretaries of both stations confirmed they did not receive calls from the Marine Rescue Co-ordination Centre (MRCC) at Shannon until 10.23 p.m. and 10.24 p.m. respectively. This was some 37 to 38 minutes after a Mayday call between Dublin Port radio and Dublin radio, part of the national VHF network.

There were also questions about the lack of Air Corps involvement until the following day. The Department of the Marine explained that it had tasked the Shannon medium-range helicopter, and had requested assistance from the RAF in Brawdy, but had decided against calling out the Air Corps Dauphin at Finner camp so that it could provide west coast cover while the Sikorsky was engaged. Significantly, a Defence Forces spokesman pointed out that there was no requirement to have 24-hour helicopter cover on the east coast since the Dauphin had been redeployed to the north-west. The Alouette at Baldonnel was only equipped for daytime cover. This was an issue that would arise again on the east coast several years later.

In all, 11 of the *Kilkenny*'s crew were rescued within the first half-hour, most of them by the *Leinster*, and taken immediately to the Mater and Beaumont hospitals on Dublin's northside, while an urgent request was put in to provide members of the Dublin fire brigade with metal-cutting equipment. In all the commotion, knocking had been heard from the ship's hull. As the search for three missing crew continued through the night, with assistance from the Garda Sub-Aqua team in the early hours, there was mounting concern about the ship's container cargo.

Part of the consignment included methyl acrylate and a resin solution for the Asahi textiles plant in County Mayo. An 'acrid' smell was reported from the scene by a member of the rescue team. Floating containers made for a navigational nightmare in the immediate area, and Dublin Port tugboats and the Howth lifeboat worked continuously to locate loose containers, some of which were towed in to the North Wall extension.

The following day the bodies of two of the three missing crew were recovered. Patrick Kehoe (49), a single man from Wexford town, was first officer on board the *Kilkenny*. He had served as a ship's master with Irish Shipping and had joined the B & I Line in the mid-1980s. David Harding (54) from Howth was bosun. A father of three, he had worked formerly as a fisherman and had only recently transferred to merchant vessels. There was no sign of the body of Desmond Hayes (29) from North Strand, Dublin, the only son of Dessie and Ann Hayes. He had been at sea since he was a teenager and was highly thought of by those who had worked with him; he had last been seen alerting colleagues to the fact that the ship was in trouble.

Captain Liam Kirwan, director of the Irish Marine Emergency Service, estimated that the salvage could take 20 to 25 working days, weather permitting, and the first task was to secure the ship which had been holed from the top of its gunwale to the bottom of its bilges on the port side. Meanwhile baby buggies and disposable nappies, believed to be part of the ship's cargo, were washed up at Rush on the north Dublin coast.

Bad weather hampered salvage efforts, but by late December most of the ship's 123 containers had been accounted for and the last remaining hazardous cargo was recovered. Christmas passed, and a German company was awarded the contract to remove the vessel by breaking it into three sections. The bow section had been brought ashore, and the midship and stern sections were still in the bay and awaiting removal, when events took a strange and tragic turn. Solicitors for Donal Gawley, a retired project manager from Carrickfergus, Co. Antrim, confirmed to *The Irish Times* in mid-February 1992 that an injunction might be served on B & I to prevent further salvage of the ship. Their client, Mr Gawley, was concerned about the fate of a container carrying the ashes of his son from South Africa to Ireland.

B & I's response to the whereabouts of their son's cremated body had caused Mr Gawley and his wife and family considerable anguish.

The Gawleys had lived in South Africa for a number of years and their son Michael was killed in an accident there in 1983. When the couple retired back to Ireland, they arranged to ship furniture and personal goods, including a casket containing Michael's ashes. Their belongings were on the *Kilkenny* when it was involved in the collision in Dublin Bay. The couple said that B & I had given vague and unsatisfactory information on the container's location, and had referred the couple to their insurers. Mr Gawley was worried that the container could still be on board the vessel, perhaps in the midship section, and that it could be damaged during the salvage work.[1]

The injunction was never served, as B & I responded. Just over a week later, on 21 February, Mr Gawley found himself speechless when trying to talk to this *Irish Times* reporter. The missing container had been located by Liffey Marine Ltd, commissioned to carry out the search by B & I Line. The casket containing his son's ashes had been found in the drawer of a sewing machine cabinet.

THE *KILKENNY/HASSELWERDER* COLLISION was one of the first night taskings for the new Shannon medium-range helicopter which was already proving its worth on the west coast. A month after the collision four seamen died when a Soviet fish factory ship, the 240 foot *Kartli*, hit heavy seas off the Scottish west coast in a westerly gale. Two RAF Wessex helicopters from Aldergrove in Belfast were involved in the rescue of 51 crew, which was co-ordinated by Clyde Coastguard. A police interpreter who spoke to the survivors at Altnagelvin Hospital in Derry said a freak wave had smashed through the wheelhouse windows, killing two crew instantly and injuring two, both of whom died later. Marine experts expressed considerable surprise that wave damage could be sustained by a vessel of that size.

Noel Donnelly, one of the most experienced Irish search and rescue winching crew, is not a man who would ever take the Atlantic for granted. He served nine years with the Air Corps, mostly on search and rescue with both the Alouette and Puma, and left in 1984 just before the arrival of the new Dauphin fleet to take up a job as a diver with a civil engineering firm in Aberdeen, Scotland. Subsequently he worked as an ambulance driver for a time in County Kildare.

He was on duty at the new Shannon search and rescue helicopter base on 1 February 1992 when the crew received a call-out at about 1 a.m. The 32 foot *Ocean Tramp*, a charter boat for sea angling trips, was on a delivery run. The vessel had three crew on board, Austin Gill, and Tom and John Groden. The three didn't know the area and thought they were east of the Aran island of Inis Oirr when they ran up on rocks in Gregory Sound. They were, in fact, between two islands on the eastern side of Inis Meáin under 40 to 50 foot high cliffs.

'Al Lockey was pilot, Nick Gribble co-pilot and Ian Harris was winchman,' Donnelly remembers. 'We searched east of Inis Oirr, based on the information given to us, and couldn't find anything. Then, as we were flying towards Inis Meáin we spotted a masthead light. The three men were up on the bow, which was on the rocks, and they were being bashed about by the waves. Ian Harris was ex-Royal Navy with loads of experience, and he went up and down three times to get them. He had taken two of them off, and one guy fell into the water. At this point the boat was virtually submerged, and so Harris had to grab him and winch him out of the sea. Fortunately the weather was reasonable, with a one to two metre sea.'

The rescue barely made it into any of the newspapers the following day, due to stormy events ashore — a heave against the Taoiseach, Charles Haughey, by supporters of finance minister, Albert Reynolds.

THE THREE MEN ON the *Ocean Tramp* were lucky to survive. Fortune did not look so kindly on the three crew of a south-east fishing vessel involved in a tragic accident later that year, while the crew of an RAF Sea King called out to save them had a close call. It was about 1 a.m. on 21 September 1992 when the 56 foot *Orchidée*, a wooden-built fishing vessel originally built in France and owned by Jimmy Power of Dunmore East, Co. Waterford, put to sea. It was heading for the 'Smalls', a fishing ground off the Welsh coast.

The crew on board were the 28-year-old skipper and owner Jimmy Power, and Robert (Bobby) Doran (19) and Kenneth Pierce (23), both of Bridgetown, Co. Wexford. The vessel arrived at the grounds around 7 a.m. that day and began fishing. It was only at about 5 p.m. that the gear was finally hauled in and the catch sorted, cleaned and stowed with ice below decks. The skipper decided to lie-to for the night and set the gear again the following morning at first light.

At about 1 a.m. Ken Pierce relieved Bobby Doran on watch in the wheelhouse. The vessel's 'not under command' lights — two red lights vertically displayed — had been switched on. At about 3.30 a.m. Pierce looked at the radar and saw two targets. One was about one and a half miles away on the port bow and the other about a mile astern of the *Orchidée*.

Seconds later Pierce saw two white lights loom up on the port bow, and the *Orchidée* was hit. He tried to reach the engine control to put the vessel astern, but it was too late. He yelled to the other two crewmen down below; they clambered out of their bunks and came up, and Doran issued a Mayday distress call on the VHF radio. Pierce climbed up to the top of the wheelhouse to try and release the liferaft, but there was a mast lying on top of it. Skipper Power called him back down, gave him a lifejacket, and all three crew with their lifejackets on jumped into the water. The weather was about force six to seven, with waves breaking over them.

Afterwards Pierce described how the lights of the *Orchidée* had gone out on impact and they were in total darkness, with a 12 foot swell running. He lost contact with his colleagues, but heard the skipper say he was going back for Doran who couldn't swim. Power headed for the stern of the vessel — and that was the last Pierce heard of him alive.

Pierce was 30 minutes in the water when he was rescued by a French fishing vessel, the *Agena*. The 95 foot steel vessel had struck the *Orchidée* on its port side with such impact that the Irish vessel had been thrown 180 degrees and collided with the port side of the *Agena* before sinking.

The accident happened 50 miles off the south-west Irish coast in international waters. Many vessels, French, British and Irish, heard the Mayday and steamed to assist. The *Agena* steamed back and forth over the area seeking the two missing crew, but there was only scattered debris on the water and a strong smell of diesel.

The official investigation stated that calls were heard from both the port and starboard side of the *Agena* as it was recovering Pierce from the water. A transmission beacon was thrown in the direction of the second voice, and it was then picked up by the vessel's searchlight. There was nothing to be seen on the water, and no further sound.

Three helicopters were called out: an RAF Sea King was tasked from Brawdy, as was the Shannon medium-range helicopter which was airborne at 4.15 a.m. A Royal Navy Sea King from the British military ship, HMS *Cumberland*, was also dispatched as the ship was in the area and co-ordinated the search throughout that day.

As dawn approached, that search also involved the Naval Service patrol ship, LE *Orla*, the Rosslare lifeboat and up to a dozen fishing vessels. The RAF Sea King, under the command of Squadron Leader Dane Crosby and co-pilot Flight Lieutenant Steve Johnson, was about two-thirds of the way through its search, and the crew were talking about refuelling, when it ran into trouble.

'I remember we had been out there for about an hour and a half at least, and had just taken our night vision goggles off. We were discussing whether to refuel back at base or on the *Cumberland* and had recently made contact to find out where the ship was, when we got a signal that the gearbox had run out of oil. There was an oil leak. We realised fairly quickly that we'd need to land the helicopter on the water, as once the gearbox packed up we would take on the flying characteristics of a brick. At first we thought we might make it to the *Cumberland*, but then the situation got worse. So we went into a low hover, and the two guys in the back jumped out. They had their individual dinghies strapped on their backs, and they also took out a multi-seat dinghy which we had on board. The sea state was very reasonable, at about three, so we backed off them and landed close by. We then shut down both engines, waited for the rotors to stop, inflated the flotation bags on the aircraft and exited through the windows. We then swam away from the aircraft in case it turned over on us,' Johnson recalls.

'We also had dinghies strapped under us. I think we were waiting about three minutes to be picked up. In fact I had just got into my dinghy and had thrown out the sea anchor when the captain was rescued. Myself and Crosby were taken by the Royal Navy Sea King, and the two crewmen were rescued by the Shannon Sikorsky. We were all flown to the *Cumberland* and given showers and breakfast. A second Sea King from Brawdy, flown by Rocky Boulden, was called out to continue the search, and he then came and collected us on his way back. The helicopter rolled over about 15 minutes after we had left it — it is very top heavy — but Royal Navy divers attached flotation bags to it to make sure it didn't sink. A ship came out the next day from Milford Haven and towed it back in.' (At time of writing, the £6 million sterling helicopter is still flying.)

Ditching procedures form a vital part of the training for search and rescue crews. At the Robert Gordon Institute in Aberdeen, Scotland, where RAF crews are sent, a simulated helicopter cabin is dropped into water and the crew have to carry out drills — 'brace . . . mark the escape exit . . . wait seven seconds . . . release the harness . . . escape to surface . . . climb into the dinghy. To make it more effective, the simulated emergency also involves waves, wind and artificial rain.[2] 'It is not something that happens very often, thank goodness,' Johnson says. 'But the important thing to remember is to have two or three helicopters in the area when it does!'

Hopes for the survival of Jimmy Power and Bobby Doran faded as the search continued and the hours went by. Power, married with two children, was a member of the Dunmore East Fishermen's Co-op. He had recently put his vessel up for sale. Doran was the son of a County Wexford fisherman, Ger Doran.

In a subsequent court hearing in France, Tracey Power, Jimmy's widow, described how he had been a fun-loving father to their two children, Jenna and Kevin, and his dream from the time he was eight or nine years old had been to own his own fishing vessel. 'He was well liked in the fishing community, a loving husband and my best friend.'[3]

At a full judicial investigation carried out by the French authorities, a hearing at the Lorient maritime and commercial court on 27 September 1996 found the officer on watch and the watchkeeping rating on the *Agena* to be guilty of negligence, causing the loss of the *Orchidée* and the deaths of two crewmen. They were sentenced to terms of imprisonment which were suspended by the judge. A subsequent report published by the Minister for the Marine and Natural Resources, Dr Michael Woods, found that the collision occurred because the *Agena* had not complied with the international regulations on prevention of collisions at sea. Both vessels had been apparently well furnished with navigational equipment which was in good working order, and the bridge/wheelhouse on both vessels was manned, the report said. 'Both apparently saw each other before the collision but took no action to prevent it.'

Minister of State and TD for Wexford, Hugh Byrne, said he was 'disappointed' that the report did not satisfy the bereaved families. He asked his department's officials to meet the families to assure them that the report did not apportion blame. The department denied it had come under pressure to withdraw the report in the light of Minister Byrne's reaction.

THERE WAS ANOTHER CLOSE call with a rescue helicopter that same year: this time it involved the Air Corps. Captain (later Commandant) Dave Sparrow, originally from an army family in the

Curragh, Co. Kildare, was on duty at the time in Finner camp. Sparrow was one of a generation of 'top guns' in the search and rescue squadron at the time, having gained his wings in 1979 and flown helicopters from 1982. There was no tasking that was too much of a challenge for him or his colleagues.

On 7 November 1992 the Finner helicopter was asked to airlift an injured man off a fishing vessel west of Blacksod in County Mayo. 'It was a murky, foggy night, and so we decided to take full fuel, land at Blacksod and fuel up again,' Sparrow, now flying with Cityjet commercial airline, remembers. His co-pilot was Captain Paul Hayes.

'We flew off the coast all the way down to Mayo, and I was in constant contact with Vincent Sweeney, attendant keeper at Blacksod lighthouse, because I was worried about the fog. Vincent was as reassuring as ever — he'd always tell you he had the kettle on and the Hobnob biscuits ready, because there were situations where we were kept waiting at the Mayo lighthouse until we received further instructions. He said there was no fog at Blacksod.

'But five minutes before we were due to land there, he told us the fog had rolled in from the sea. We were worried that if we went out to the casualty, we could be caught with landing and refuelling on the way back. So we tried two approaches to Blacksod and couldn't see it.

'We overshot it and set off to return to Finner. An RAF Nimrod had been called to give us top cover, but as he was flying down towards us, he noticed that the weather was good at Finner, so he headed back to Lossiemouth. However, five minutes before we were due to land there, the fog rolled in.

'We thought about Killybegs, then tried Sligo Airport, but conditions were too bad. I considered ditching beside Inishmurray Island, just off Sligo. I briefed the crew and said we'd transition down to 50 feet, put on our survival gear and the crew would jump out. Then I'd point the helicopter to sea and jump out myself. We discussed it among ourselves and thought that perhaps the fog might only be on the coast. So we decided to try flying inland, and one of the crewmen was first to see lights. We landed in a little tiny field in a place called Coolaney in Sligo at 4.30 a.m.

'We climbed out and gave each other high fives, and one of the crew kissed the ground! We were all geared up in our immersion suits and Mae Wests [life vests] and we looked like spacemen. Next thing this dog appeared and jumped on me; if I had had a gun, I'd have shot it. We fought the dog off, and I asked the lads to go and find a Garda station.

'The two crewmen, Dick O'Sullivan and John McCartney, set off up the town and came back 20 minutes later with large grins on their faces. They told us to come with them, which we did, and they brought us to the town's only pub. We couldn't believe it; it was called Happy Landing.

'Unfortunately it was closed, but we went to it the following morning and had our photographs taken outside. I returned there a year ago with my wife Mary and our kids, and the lady behind the bar recognised me. She brought out one of the photos, sure enough! I asked her then how it got its name. Apparently some guy had built an aeroplane and got lost in flight. He ended up landing in a field close to the pub!'

The crew had also landed close to the home of Joe Corcoran, air-traffic controller at Sligo Airport. 'Joe was sitting in the control tower in Sligo making weather observations in the event that we would attempt to land there,' Paul Hayes says. 'When he heard we had landed at Coolaney, he got into his car and went home to put on a slap-up fry for us. Of all the agencies involved that night, Joe's service had to be the tastiest!'

'We got a lucky break, just like the first guy after whom the pub was named,' Sparrow observes. 'After the Dauphin helicopter crash in Tramore seven years later, when conditions were similar with that sea fog, it brought it all back to me. We could have been them. It was that sort of close-run thing.'

PURE LUCK WAS ALSO with a swimmer who got into difficulties off the Forty Foot in Dun Laoghaire on the morning of 27 March 1993. The Irish Marine Emergency Service — now the Irish Coast Guard — had just established the Marine Rescue Co-ordination Centre (MRCC) in Dublin; previously it had been at Shannon Airport. Commandant Tom O'Connor, son of one of the first Air Corps SAR pilots, Commandant Fergus O'Connor, was out on a routine training mission in the Alouette III from Baldonnel with winch crew Flight Sergeant Daithi Ó Cearbhalláin and Corporal Neil McAdam. The helicopter informed the MRCC it was going to head out to Dublin Bay and fly up to train with the Skerries lifeboat.

'We were crossing Ballsbridge when the MRCC got a call to say that a swimmer was in distress at the Forty Foot,' O'Connor remembers. 'There was a north-easterly and it was one of those crisp, cold days. We were there within seconds and came into a hover, and saw this man treading water below. Neil McAdam went down on the wire to check him out. He was very big. Just as Neil approached him the man fell unconscious. Neil spent a good two minutes trying to get him into the strop, but he was greased for his swim and only had his togs on. Neil is very strong, but the man was like dead meat. Daithi winched them up but they both had great difficulty getting him through the door — and Ó Cearbhalláin is ex-Ranger wing, so he's a tough guy.'

There was no option but to land, which O'Connor did — on the green opposite Teddy's ice cream shop in Dun Laoghaire. 'We got him into the aircraft then, flew him to hospital and he was there for four or five days.' It emerged afterwards that the man was an experienced swimmer who had just got caught in a current. Mobile phones were only emerging on the market in 1993 and few people had them; fortunately a couple who were out walking did have one, heard about the swimmer in trouble and rang the emergency services.

'That combination of events was crucial to that man's survival,' O'Connor says. 'And if we hadn't been flying out that way when the MRCC got the call, valuable minutes would have been lost.' The phone call was logged at 10.51 a.m. and the swimmer was *en route* to St Vincent's Hospital by 11.01 a.m. Afterwards an evening newspaper described the rescue as one of the 'fastest on record'.

FORTUNE WAS ALSO WITH Captain Paul Hayes when he was tasked to find two missing divers off north Donegal the following year. Locating a vessel at sea is one thing; locating an individual or individuals visually, without specialised heat-seeking equipment or a satellite navigational fix, is an almost impossible task. Until manufacturers began making luminous safety devices for divers, their black wetsuits were a dangerous form of camouflage in an emergency at sea.

Aircraft commander, Commandant Jim Corby, and Hayes knew they had their work cut out for them. The location was the treacherous north-west stretch of coastline between Tory and Inishtrahull Island in Donegal which has claimed many ships over the centuries. Like Rathlin Sound to the north-east, it is a diver's paradise, given that many of these wrecks lie in less than 35 metres of water and visibility is good.

It was the evening of 13 July 1993. The Dauphin took off from

Finner camp and flew north; time was not on their side. Light was fading and it was nightfall by the time they reached Inishtrahull. The divers had been with an accompanying vessel but were several hundred metres from the boat when they surfaced with empty tanks. They couldn't see the boat and lost their bearings. The alert was raised by the companion vessel.

'In this sort of situation the Dauphin's computerised navigational equipment helps,' Hayes, now retired from the Air Corps and flying with Aer Lingus, recalls. 'When you are travelling at 80 miles per hour and you spot something, by the time you turn it can be too late. But fix the location on the computer and you can return to it. We carried out a very lengthy search, saw nothing, and we were quite a bit offshore. I was just going back to refuel in Derry when something caught my eye. It was a shadow on the water, and I thought it might have been a seal. I logged it in the computer and, sure enough, it was the missing divers. They were a couple from Northern Ireland. We winched them on board, took them to Altnagelvin Hospital and they were in good shape. It was a fluke glimpse. If we hadn't seen them then, we might never have found them on returning back out.'

The two divers were delighted to see the Dauphin; but Hayes and his colleagues have had several experiences where those rescued aren't always quite so appreciative. One such was in 1994, the year after that Inishtrahull mission, when a group of students from NUI Galway went body boarding in Sligo Bay. Some were only wearing light neoprene suits, which wouldn't have given them much heat protection if in the water for any length of time.

Hayes was on duty at Finner when the call-out came through. The students had been swept out to sea. 'Again, it was a very lengthy search and the Sikorsky S-61 from Shannon was brought up to work with us. The Ballyglass lifeboat, only recently stationed in north Mayo, was also called out. It was the lifeboat that found them, and at this stage they were east of Killala Bay. The lifeboat crew heard them screaming — fortunately they had had the sense to stick together, but they had been carried a considerable distance south. Next day, the students were being interviewed on radio about their experience. They made light of it and said they were never in any danger, and that if they had been in any trouble they could have swum back up the coast.'

Superintendent Tony McNamara, who was coxswain of the Ballyglass lifeboat that evening, was appalled at the students' response afterwards on national radio. 'They did the right thing by tying the boards together, but it was pure pot luck that we found them with the lifeboat's spotlight. One of the girls was hypothermic, and I remember we had to strip off their wetsuits and give them some of our own clothes to get them warm. I handed my RNLI jumper to one of them, thinking I'd surely get that back. I never saw it again! If we hadn't got to them when we did, they'd have been bodies the next day.'

THE DAUPHIN'S MANY STRENGTHS — its speed and technology — were being stretched to the limit at the north-west search and rescue base. Captain Hayes was pilot on duty at Finner during one of those occasions. The weather wasn't particularly bad on 4 July 1995 when a fishing vessel, the *Radiant Way*, was reported to be on fire some 40 nautical miles north of Tory Island, Co. Donegal. Although it was close to the Irish coast, it was within the British Coastguard's area of responsibility and was handled on the day by Belfast Coastguard; the Irish marine rescue area is based on the Shannon flight information region (FIR) boundary.

On board the helicopter with Hayes were Captain Kevin Daunt, Sergeant Daithi Ó Cearbhalláin and Airman John Forrestal. The vessel was drifting, and another trawler was waiting close by. 'It was a hot summer's day with a calm wind, and our plan was to winch the crew off and fly them ashore,' Hayes recalls.

'The problem is the Dauphin's turbine engines have to work harder in hot, calm wind conditions. At 36,000 feet over the Atlantic in cold air, aircraft turbine engines tend to be more efficient. A helicopter hovering in strong wind conditions also requires less engine power, as even before the rotor blades start to turn there is wind energy flowing over the aerodynamic surfaces. We flew with minimum fuel to ensure that we had the power reserves — although the Dauphin can dump fuel if necessary. The gearbox was working at 98 per cent, I remember, and we had to pick the crew up one by one. It is also very hard to hover with limited power, so we decided to transfer the crew, singly, over to the other fishing vessel just to be on the safe side. Fortunately we managed to get them all off, but it was one of those examples where the Dauphin's limitations were exposed.'

8

HEROES UNSUNG

Quips about the 'worm' and the 'dope on the rope' belie the hazards facing winching crews. For Ben Heron, however, the 'end of the wire' was the only place he ever wanted to be during his career in the Air Corps. According to Ben, 'You're at the sharp end, dealing with all sorts of situations, and there is the most tremendous satisfaction if things go according to plan.'

Sergeant Heron wasn't on the wire, but was winch operator on duty when he received a phone call at Finner camp just before 6 p.m. on 25 September 1993. It was the ambulance controller at Ballyshannon down the road. He had received a report through the 999 emergency system that a man had fallen off a cliff at Horn Head in north Donegal, close to the old coast guard station.

Heron alerted the duty crew, Captain Dave Sparrow, co-pilot Captain Mick Ryan and winchman Airman John McCartney, and asked the ambulance controller to pass on the relevant information to the Marine Rescue Co-ordination Centre (MRCC) in Dublin. He also advised the controller to send an ambulance to Horn Head in case the helicopter crew needed assistance with the casualty.

Rescue 110 was airborne at 6 p.m. and flew directly to Horn Head. Heron called up Malin Head coast radio *en route*. The radio operator told him that the Mulroy Bay Coast and Cliff Rescue Service was on its way there. Within 30 minutes the helicopter was at the headland, and several civilians signalled the precise location; one of those on the ground told the crew that two people had fallen off the cliff. A northerly 15 to 20 knot wind was blowing at the time, with some cloud and good visibility.

The Dauphin was closing in when Heron spotted a man trapped on a very small ledge about halfway down the cliff face. The pilot asked him to 'patter' him into a position at the summit of the cliffs. As Sparrow manoeuvred out over the summit, he hit severe turbulence and ran out of power. The aircraft was forced to dive away over the sea.

The pilots decided to jettison fuel; they would need maximum power to hover, and there was about an hour and a quarter left till darkness. Sparrow reduced his fuel load to 400 kilograms. As the helicopter returned, Heron noticed a second person who appeared to be motionless among scree and boulders at the base of the cliff.

The crew agreed that Heron should jump out from a low hover and get more information from the man on the summit. He identified himself as Paul Rintoul; his 20-year-old twin brother Chris and a friend, Peter King, had 'gone over the edge of the cliff'. Chris was the man halfway down on the ledge, and Paul Rintoul wasn't able to see King.

Heron climbed back on board the helicopter, briefed the crew, and the Dauphin flew a circuit and approached to carry out a detailed reconnaissance of the situation. The helicopter flew in towards the ledge, and they could see that Chris Rintoul was in serious trouble. As Sparrow noted in his mission report, 'I could scarcely believe my eyes.'

It was, Sparrow recalls, 'something you'd see in a Roadrunner cartoon, where a character is in the most impossible situation.' Rintoul had his back pinned against the ledge and was supporting himself with one foot on a stone, his heels dug into the cliff face, while clinging to several bits of scrub grass on either side of him. Directly below him was a 250–300 foot sheer drop. 'He was totally unable to move and could only nod his acknowledgment of our presence.' Unfortunately he could also see the lifeless figure of his best friend Peter King down below.

The aircrew decided to attempt to lift Rintoul off first, as he appeared to be very weak. There was also the risk that if he fell while the helicopter was attending to King, he would almost certainly hit the aircraft — with horrible consequences. The cliff was so sheer and the ledge so small that Sparrow had to climb about 90 feet above Rintoul to position overhead.

The crew discussed the plan: Heron would winch McCartney to a position level with and just to the right of Rintoul. This 'position' was about three foot square, two yards to the right of him. McCartney would have to stay attached to the hoist cable at all times; there was the possibility that the helicopter might fall away, due to turbulence and lack of power.

'For John, it was a hell of a leap of faith,' Sparrow says of his winchman. The pilot climbed to position the aircraft some 100 feet above Rintoul, with the co-pilot, winch operator and winchman providing constant patter on the blade and tail clearance from the cliffs to port and directly in front of the helicopter. Sparrow found himself using 'unnatural control positions' to maintain a steady height. McCartney was winched out and Heron managed to land him on the earmarked patch, but the winchman couldn't get a firm footing. Clinging to a clump of heather with his right hand, McCartney reached across with his left hand to try and put the strop around Rintoul's arm. Rintoul was terrified. With great patience and courage, McCartney managed to inch the strop on to one arm and partially behind his head.

Back on the aircraft the pressure was almost unbearable. 'One mistake would certainly pull him from the ledge,' Sparrow remembers. 'And we were hovering so close that the radar radome on the front of the Dauphin was virtually touching the rock face, the blades were clearing the cliff top in front by only a few feet and were just four to five feet from the cliff to port.' As he wrote in his mission report, 'The whole aircraft was vibrating in the turbulence and the flight controls and power were at the limits of travel.'[1]

Would the scrub and heather hold? The winchman prayed it would as he kept shouting to Rintoul, trying to reassure him. Once the strop was around Rintoul, he then had to persuade him to sit forward to place it correctly and tighten the toggle. Naturally, Rintoul was very reluctant to risk this; he would loosen his grip. McCartney swung his leg across him to help him feel more secure, and it did the trick.

The winch operator was working hard to maintain the correct tension on the cable, while the pilots were relying on Heron to keep them informed with his constant patter. Once McCartney gave the 'thumbs up' signal, Heron 'snap lifted' both winchman and survivor from the ledge. The sudden additional weight of both at the end of the cable forced the helicopter to career away to starboard, and it fell downwards from the cliffs over the sea.

'At this point Rintoul became quite distressed,' Heron recorded. 'Airman McCartney continued to reassure him, and by holding the survivor's head close to his own body he prevented him from seeing just how high above the sea they were.'[2] Once clear of the cliffs, Heron winched them both on board at about 200 feet. Both he and McCartney checked Rintoul for physical injuries, found he had none, but realised he was in a very shocked state. They directed the pilots to land beside the waiting ambulance and have him taken directly to hospital in Letterkenny.

The winch crew were only minutes on the ground, dealing with the ambulance crew and local gardaí, before climbing back into the Dauphin and taking off again to recover Peter King. The helicopter flew down towards the cliff base and Heron winched McCartney out with a stretcher on to the rocks below. King was lying on a large, flat boulder. McCartney needed to assess his medical condition, so the helicopter flew out to sea and did a few circuits.

King showed no life signs. McCartney lifted him into his stretcher and set off a yellow smoke flare to let the Dauphin know he was ready to lift. It wasn't the best winching location, and he couldn't move the casualty. The helicopter climbed about 80 feet and moved in quite close to the cliff to reach him. Heron lifted both McCartney and the stretcher bearing King clear, and the Dauphin flew out again over the water and descended. Heron recovered both of them on board at about 50 feet.

The Dauphin advised Malin coast radio and flew to Fanad Head to refuel. It landed at 7.35 p.m. and was airborne again in 20 minutes *en route* to Letterkenny Hospital. Sadly, King was showing no life signs at all. He was pronounced dead on arrival by the doctor on duty in the hospital's casualty department. The Dauphin returned to Finner, landing at 9.15 p.m.

A month later the Defence Forces received a letter from Alan Rintoul in Belfast. He had written to the General Officer Commanding the Air Corps 'to commend the bravery and skill of the helicopter crew' for their rescue of his son, a student at Queen's University.

'My son, Christopher, owes his life to this team, and I would like to pass on my sincere thanks and appreciation from all my family for their actions on that day', he wrote. 'Captain David Sparrow, Captain Michael Ryan, Sergeant Ben Heron and Airman John McCartney have been commended to me by everyone who saw their actions on that fateful day. Indeed, talking to local people and the Garda, it is quite clear that they put their own lives at risk and performed one of the most dangerous rescues that any of them had seen or heard of before. I consider that you are a very fortunate man to be able to command such dedicated professionals.' Mr Rintoul added that Raymond King, father of the young man who died, 'asked to be associated with these comments and to personally thank you for your efforts to save his son.'[3]

Ten years after the rescue, Alan Rintoul still wonders why the aircrew never received an award for their efforts. His own father was one of the first helicopter engineers to train with the RAF. 'So I do have some knowledge of the efforts that the Dauphin crew made that day, and in an aircraft which was not designed for those conditions.' Significantly, his other twin son, Paul, who raised the alarm, had been over the other side of the head when the accident happened. 'He was with my dog, and for some reason he sensed that something was amiss and returned.' Mr Rintoul attended the inquest into the death of Peter King. 'I remember the winchman describing how he didn't sleep properly for a week afterwards. I found that most impressive — that he could show such courage, and yet admit this also.'

The pilots and winch operator were also impressed with the efforts of their colleague, Airman McCartney. 'As an experienced operator, I have been involved in several difficult missions, but I have never seen a mission which demanded such professionalism and courage from a winchman as was demanded of Airman McCartney on this job', Heron wrote in his report.[4] And Dave Sparrow said that McCartney had displayed 'an exceptional level of determination, courage and skill, coupled with an apparent total disregard for his own life in the execution of this hazardous mission.'[5]

SERGEANT HERON (LATER FLIGHT Sergeant and now retired from the Air Corps) was one of the many winch crew personnel who undertook medical training at the National Ambulance Training School in the Phoenix Park, Dublin, which was overseen by the Pre-Hospital Emergency Care Council. However, he had been in enough situations to know that paramedic training would be to his benefit — and that of many victims. Eamonn Burns, formerly of the Air Corps and now with CHC Helicopters at the Irish Coast Guard base in Shannon, travelled to the US to do paramedic training. While attending the North Eastern University, he suggested that the college authorities should run such courses back in Ireland.

'I did that course, and also completed instructor's training at my own expense,' Heron says. 'One gets to know how to use a defibulator and other vital medical equipment which can assist in treating a casualty en route to hospital.' Unfortunately, at time of writing, the qualification was still not officially recognised for either helicopter or ambulance crews; they are not allowed to break the skin or administer drugs during an emergency. It is a situation which medical practitioners like Dr Marion Broderick on the Aran Islands would wish to see addressed at official level.

There is a very practical reason for recognising the qualification, Heron points out. 'You take a situation on an air-ambulance run where you have a medical crew with the patient. That doctor and nurse may be excellent at their job, but they haven't been trained to work in helicopters, and in such a noisy and constantly vibrating environment it can be very difficult to detect by stethoscope if someone is breathing or to take a pulse. There have been situations where the turbulence is so bad in the air that the medical crews themselves get sick and that can put the winch crew under severe pressure in the back.'

Heron recalls one air-ambulance mission during his time in the Air Corps that did upset him greatly. It was just a few months before the Horn Head rescue. On 5 July 1993 some children were playing in the driveway of a house in County Donegal when the handbrake of a parked car was left off. A three-year-old child was crushed between the car and a wall and was rushed to Letterkenny Hospital. The toddler was seriously injured and required urgent treatment at Our Lady's Hospital for Sick Children in Crumlin, Dublin.

The Finner helicopter was called out to fly the girl to Dublin. 'I remember that my own first kid was about the same age,' Heron says. 'Captain Dave Sparrow was the pilot, and Crumlin wasn't prepared for night landings. By the time we got there, there was hardly any blood pressure but the medical team kept working on her all the way.'

Sparrow, who was flying with co-pilot Sean Murphy, recalls that he 'got into a little trouble with authority' afterwards for taking on a night approach to the hospital without clearance. 'The ambulances were set up in Baldonnel to meet us, but we went straight to Crumlin. It is the kind of thing you have to do when time is of the essence, and there wasn't much sense in landing in Baldonnel, given the little girl's condition. We had hand-held night vision goggles for the journey on the way down from Letterkenny. However, there was no night helipad at Crumlin, and so we literally picked a spot outside the intensive care unit. The winch guys hung out the door and "pattered" us down. The little girl was rushed into theatre, and we took off again with the medical team and returned to Finner.'

'We thought we had done really well to get there, and we were having a cup of tea with the medical team back at the base,' Sparrow recalls. 'There was a sort of unwritten rule that we didn't always press to find out what happened afterwards in that sort of situation. We didn't want to put people under pressure, I suppose, and also there is a sense of not wanting to know, in case it didn't work out. However, one of the nurses with us went over to use the phone, and I remember watching her face crumple. My co-pilot took off out of the room, he was so upset, and went for a long walk. We were gutted. It broke our hearts. We certainly did lose sleep over that.'

Ben Heron concurs. 'Everyone had pulled out all the stops. It knocked us for six.'

9

DUNBOY'S NEAR-DITCHING

Winds of 92 miles per hour. Mountainous Atlantic rollers. Electricity blackouts along the western seaboard. John McDermott will never forget that 24 hours, less than three weeks before Christmas 1993.

Formerly a sergeant with the Air Corps, McDermott had transferred to the rescue base at Shannon, then run by Irish Helicopters for the Irish Marine Emergency Service. 'I remember that the winds were so strong that cargo palettes were flying around the airport. We were on duty, and when we looked out of the window we could see a Fiat Uno belonging to one of the pilots lifting up and down. I was watching my own car parked near a big gate. I was worried that the gate would break and smash against the vehicle, but when we tried to go out to move it, we were forced to retreat!'

McDermott and his partner, co-pilot Carmel Kirby, finished duty about 9 p.m. and headed back out to Quin, Co. Clare, some 11 miles away from Shannon. Kirby, from Renmore, Galway city, was the first female pilot on the Shannon search and rescue unit, and had been assigned in May of the previous year. The storm was so severe that they had to make several detours. When they eventually arrived home, a front wall had collapsed, a tree was down and there was no electricity. As they lit candles in the house, they remarked that nothing would happen that night. Conditions were too bad for anyone to be out.

They were wrong. Several Spanish fishing vessels had put to sea from Irish ports in the teeth of a gale warning. Among them were two Irish-registered Spanish fishing vessels, the 35 metre *Dursey* and the 42 metre *Dunboy*, owned by the Spanish company, Eiranova Fisheries, in Castletownbere, Co. Cork. Like most of the Spanish vessels working in Irish waters, they were under pressure to earn their keep and make the most of the fresh market before Christmas. They were about 25 miles apart off the west coast when they began experiencing communication difficulties in storm force winds.

In the early hours of 9 December the *Dunboy* reported an engine failure and gave its position on VHF radio as 40 miles west of Slyne Head, Co. Galway. To the south-west of it, a third Spanish vessel, the *Mara Sul*, also reported engine difficulties about 75 miles west of Loop Head, Co. Clare.

McDermott was fast asleep when he was awoken by his bleeper at 1.55 a.m. At that time the simple call-out system gave no further information. The winds had only abated slightly and were still blowing at up to 70 miles per hour as they headed for the car and drove back in through floods to Shannon. 'We were told there were 13 crew on a Spanish boat in trouble. I remember thinking that was a great number — 13.'

The Naval Service patrol ship LE *Aisling* had picked up the alert and was heading for the *Dunboy*, but was many hours away due to the heavy seas. Shannon had been tasked by the Marine Rescue Co-ordination Centre (MRCC), along with an RAF Nimrod aircraft from Kinloss, Scotland, to provide top cover. Captain Liam Kirwan, on duty that night at the MRCC, decided to put in a request for the Air Corps and the RAF. There didn't seem to be any great need, he recalled afterwards, but his instincts told him he should.

As the Sikorsky prepared to take off at Shannon, the lanky engineer, Pat Joyce, almost lost his life. The helicopter's engines had started and the rotor brake was released. 'There is no great centrifugal force when the blades start initially,' he explains. 'In high winds the blades can hit the cockpit or the tail during that start-up phase. I could see that starting to happen and had to get out of the way.'

The pilot, Nick Gribble, who had left a British helicopter charter company to join Irish Helicopters in September 1991, was normally pretty jovial. There was usually a bit of banter as the crew set off on a call-out to keep spirits up. 'I was trying to crack a few jokes, but got no response this time from the cockpit,' McDermott says. The atmosphere on board was pretty tense.

The 13 crew of the *Dunboy* were up on deck, lifejackets all secured, when the Sikorsky arrived overhead. However, co-pilot Carmel Kirby could only spot an intermittent single light. 'We thought at first that the light was going on and off,' McDermott remembers. 'In fact, the seas were so bad that the vessel was just disappearing behind the swell.' As the Sikorsky tried to hover overhead, the radio altimeter showed that waves were between 60 and 80 feet.

'The waves were just heaped up behind each other, like the film set for *The Perfect Storm*,' McDermott recalls. 'I'd never seen anything like it before, and haven't since. We tried to put the hi-line on the boat and it broke, and it took 25 minutes to secure it. I don't know how I got down on deck but I did. At one point I remember landing on the wheelhouse and the radar reflector and aerial crashing up around me, and then I was whipped back up again.'

He had to organise three crew at a time to winch off, and remembers the cable burning through his gloves. 'We were all falling all over the deck, it was so bad. I got three off, and then I remember another three came towards me and one walked away before I had a chance to put on the strop. It was so wet that the radio had stopped working, and I couldn't hear myself talk. I had just got one guy into the strop when the boat listed 70 degrees. A lot of slack cable got caught up on the deck and the guy was dragged away from me. I remember I grabbed him by the ankles, and he was pulled right over to the other side of the deck. It was a miracle he didn't go over the side. At this stage, about 120 foot of cable had wrapped around the helicopter blades, but I only found that out afterwards. All I had seen was a flash as the cable broke, and I didn't know how.'

Normally when a cable breaks, the winch operator and winchman can use a splice or a hook assembly. 'So I thought they'd fly around and come back and we'd fit this. Instead, I saw the helicopter descend to within 20 feet of the crest of the waves. If it had gone in, there is no way they would have survived. The sea would have just swallowed them up.'

It was every helicopter pilot's nightmare. On board, Kirby heard a bang, looked round and couldn't see winch operator Peter Leonard. She thought at first he had fallen out. In fact he had been hit by the rebounding cable, knocking him back into the cabin, before the cable sliced into four of the five rotor blades. 'I heard the captain roaring to Peter to winch in. I put out a Mayday call because I believed we were going to ditch,' says Kirby.

'It all happened in a flash,' Peter Leonard remembers. 'It had taken two to three attempts to get John on the deck. Because of the conditions we were trying to winch up as many as we could as quickly as possible. We were trying to get the crew up on to the bow of the ship, and John managed to get three into the strops and I winched them in without any hassle.

'I put the strops down a second time, but they were very slow in coming out and so we picked up only two at the second attempt. When I winched down a third time, one of the crew was waiting for the other two to come forward, and then all of a sudden the ship lurched away from us. The winch got caught and ripped off just above the hook at the deck end.

'All I heard was a whooshing noise, and I didn't know much about it until I came to my senses a few seconds later. My headphones had become unplugged, and I ended up in front of the winch operator's seat on the other side of the cabin. I realised I had cut my nose. I looked at the winch and all the wire had come off. That 80 feet of cable had snapped from the ship's end and with the tension it sprang back towards the aircraft and wrapped itself around the rotors and ripped itself off from the hoist.

'I heard Carmel putting out a Mayday, and Nick Gribble began flying away. We knew there had been damage, but there was no visual warning in the cockpit about damage to the blades. It was only when we had shut down that we saw the extent of the damage. It was quite horrendous.'

The MRCC picked up the alert and relayed it. Out on the Aran island of Inis Mór, Dr Marion Broderick, medical officer with the Aran Island lifeboat, and Coley Hernon, Aer Arann airport manager, were asked to prepare for a possible crash landing on the airstrip at Cill Éine, but there was no electricity on the islands due to the storm. Fortunately the Air Corps and the RAF were *en route*, and the lifeboat from the Aran Islands had been alerted. The *Aisling* was still battling seas some hours away.

'All I can remember is a slight descent, seeing a bit of Atlantic, and then flying away,' Gribble told *The Irish Times* the following day. 'My primary consideration was getting the aircraft into a safe flight regime, whether in the air or on the water. It's really not a major thing unless the vibration gets too bad, and it wasn't. We were only doing our job, but we were very lucky not to have ditched.'

The professional 'speak' about safe flight regimes belied the drama of the situation. McDermott saw the helicopter fly off and then vanish — he wasn't sure what had happened to it. 'I thought it was gone, I had no radio and I was on this vessel with absolutely no power on board.' He tried to make his way into the wheelhouse and recalls that all the windows had been smashed and the metalwork was bent and twisted by the force of the waves. 'The skipper was pretty calm and told me the vessel wasn't going to sink. A number of guys below deck were just holding on to a bar. The boat was broached to the sea, it was pitch dark, and I was violently ill.'

Lying on the wheelhouse floor, he took his helmet off for some relief. However, he got cold and put it back on again. Seconds later he heard the crash of metal behind him. A fire extinguisher had broken loose in the pounding and hit him on the back of the head. 'There was a crack right across the back of the helmet. If I hadn't been wearing it . . .'

Shortly afterwards another Spanish vessel came within a quarter of a mile, and very soon it was within earshot. 'There was a lot of shouting in Spanish, and I then witnessed one of the finest acts of seamanship I have ever seen. The skipper of the other vessel had a hand-held radio which he bound up in bubble wrap. As the vessels passed in the swell, he made several attempts and then threw it over.'

McDermott remembers grabbing the radio when he heard 'Rescue 51' — it was the RAF Nimrod. 'I asked the pilot where the S-61 [Sikorsky] was, and was told it was ten minutes from Galway.' Like the Aran Islands, the city was also in darkness, the storm having felled electricity cables throughout the county. However, as the helicopter flew east, power was restored and it landed safely at Galway Airport with the five crewmen from the *Dunboy*. The pale face of co-pilot Carmel Kirby, photographed shortly afterwards, spoke volumes in the newspapers the next morning.

Back out at sea, an Air Corps Dauphin flown by Captain Donal Scanlon from Finner camp had refuelled at Blacksod and was approaching the *Dunboy*. McDermott knew the winch operator, Sergeant Dick O'Sullivan, having been trained by him. The Dauphin did not have the hover capabilities of the Sikorsky and there was difficulty getting the hi-line on board. Three times it broke. At one point the deck was swamped with the lights of the helicopter. 'It seemed as if it was going to crash down on top of us,' McDermott remembers. 'As soon as they came round again, I just waved them away. It was too risky after what had happened before.'

Hours passed. McDermott remembers watching a lone seagull which hovered constantly at the back of the boat but never landed. He was sick several more times, but felt no fear. He thought of family at home 'completely unaware of what was going on'. At one point a fisherman approached him and showed him photographs of his family. 'It was a bit sobering. I didn't have pictures of my kids with me. I got the feeling he thought this was the end.' Yet the second fishing vessel continued to stand close by, which was comforting.

At about 6.30 a.m., over an hour before daybreak, the wind died. 'It was like the flick of a switch,' McDermott says. The MRCC had informed him that the RAF was *en route*, and he consulted with the skipper and crew. They decided that they wanted to stay with the vessel and take a tow. 'I had to reassure the MRCC that there was adequate survival equipment on board.'

And so, after all that effort, the winchman was taken up by the RAF Sea King on his own and flown to Galway, where he joined the rest of his very shocked crew. That night the helicopter crew had their Christmas dinner. 'I didn't have any aftershock,' McDermott recalls. 'It hit Carmel several days later.'

The *Dunboy* was taken under tow to Killybegs, Co. Donegal. Further south, the *Mara Sul*, also in difficulties, was offered a tow by another vessel and both headed for Spain. The RAF Sea King remained at Shannon on standby until a relief Sikorsky could be flown to the base.

There were tributes from all quarters. The Minister for Defence and the Marine, David Andrews, and his junior minister, Gerry O'Sullivan (now deceased), sent congratulations to the rescue units, as did various fishing industry representatives. It was the Sikorsky helicopter's 302nd mission since the service had been established under contract to Irish Helicopters by the Department of the Marine in July 1991.

Joan McGinley, founder of the West Coast Search and Rescue Action Committee, also paid tribute to the rescue crews, but was very critical of the decision by the Spanish fishing vessels to put to sea in spite of storm warnings. She called for penalties to be imposed on such vessels, as such 'irresponsible action' places the lives of rescue service personnel at risk. The fishing skippers had been aware of the severe weather conditions forecast since the previous Sunday and yet had chosen to ignore this information, she emphasised. 'The State has invested heavily in improving search and rescue services, and we on shore should help to prevent accidents by observing weather forecasts. There should be a severe levy on boats which put out in such conditions.'

She had good reason to be angry. Just three years before, the Naval Service lost its first crewman on duty during a rescue operation involving a Spanish vessel which had also chosen to disregard an unfavourable weather forecast. Leading Seaman Michael Quinn was serving on the patrol ship LE *Deirdre* on 30 January 1990 when it responded to a distress call from the *Nuestra Senora de Gardotza*, a British-registered Spanish fishing vessel, with 17 crew on board.

The vessel had gone aground in force eight winds on rocks close to Roan Carrigbeg, off Bere Island in west Cork's Bantry Bay. By sheer chance the *Deirdre* was in the bay. It was lying at anchor off Bere Island in Laurence Cove when it picked up the distress message shortly after 9 p.m. from Valentia coast radio station. The fishing vessel had given the coast radio station an incorrect position initially, but a quick-witted lightkeeper, Nicholas Halpin, spotted the vessel in difficulty from Roancarrig Mór lighthouse.

The patrol ship was unable to establish radio contact with the casualty, and it was decided to launch a Gemini inflatable. Even as the crew were being selected, Leading Seaman Michael Quinn, a 27-year-old single man from Drogheda, Co. Louth, volunteered. He was appointed coxswain, along with Able Seaman Paul Kellett in bow. The inflatable was launched just after 10 p.m. with instructions to return to the ship if conditions proved too difficult.

Back at Roancarrig Mór, Nicholas Halpin was lighting flares to guide the Navy crewmen to the vessel. The inflatable made its way to the rocks in pitch darkness and screaming winds. They spotted the fishing vessel, but Quinn realised that sea conditions and shallow water would make an evacuation very difficult. He was bringing the Gemini round to return to the *Deirdre* when a large wave caught the inflatable, capsized it and threw both men into the sea. Both were wearing flotation suits, but conditions were so atrocious that they soon lost sight of the Gemini and became separated.

Kellett was fortunate to be washed ashore. Even as he reached land he was swept off again by another wave, but managed to scramble on to rocks again. Shocked and bruised, he hauled himself across several fields and flagged down a Garda patrol car. Afterwards Kellett described how he had seen Quinn close to him in the water at one stage, but wasn't able to make physical contact with him.

Michael Quinn didn't survive; his body was recovered by the Air Corps early the following day. The crew on board the fishing vessel were airlifted off at 2.30 a.m. by an RAF Sea King helicopter from Brawdy in Wales during an intense 24 hours for the British air-sea rescue services. Over on the east coast, the air wing had been involved in an alert when the Sealink ferry, *St Columba*, reported a fire in its engine room while on passage across the Irish Sea. The ship had 199 passengers on board, all of whom were called to muster stations and told to put on lifejackets as helicopters flew firemen to the vessel. After about two hours the fire was extinguished and a potential evacuation was called off.

There were some questions as to why the Air Corps Dauphin

helicopter — by then based at Shannon on the west coast pending the medium-range contract — had not been tasked to assist in the emergency in Bantry Bay. A spokesman for the Defence Forces told reporters that the short-range helicopter would only have been capable of carrying four people at a time and would have had to refuel. The RAF Sea King would have been there just ten minutes after it and would have been able to lift off all 17 crew.

Joan McGinley hit out at the decision not to send the Air Corps immediately to assist the RAF, and said that it added focus to the debate on the need for a medium or long-range helicopter service. The government was still 'foot dragging' on the issue, she said, while it had spent £1.3 million leasing an additional jet for the duration of Ireland's presidency of the European Union. 'One can understand that the RAF Sea King had to be called to airlift 17 Spanish crewmen,' she told *The Irish Times* on 31 January 1990, 'but the incident occurred on the coastline within two miles of a helipad and refuelling base [at Castletownbere], and we now know that the lives of two naval ratings were also at stake. When it is so experienced, why couldn't the Air Corps have been there to initiate the rescue?' she asked.

The fishing vessel crew — two British and 15 Spanish — were flown home to La Coruna and Plymouth unharmed. The local Spanish agent, Derry O'Donovan, denied that the vessel had put to sea in spite of a gale warning, saying it had been sheltering in Berehaven and was *en route* to Bantry when its steering failed. Yet as local fishermen in west Cork pointed out, the *Nuestra Senora de Gardotza* was the fourth fishing vessel to have run aground in Bantry Bay in that fortnight. Ironically, this vessel was well known to the Naval Service: it had been detained on six separate occasions for alleged illegal fishing offences in Irish waters, the last offence being in 1985.

The Naval Service said it was initiating an inquiry into the death of Leading Seaman Quinn. It emerged that he was engaged to be married to Sarah Buckley. She attended a ceremony hosted by the Spanish Government some months later, along with Michael's two sisters, Angela and Kathleen, to pay tribute to the two Navy seamen. The King of Spain's award of the Spanish Cross of Naval Merit was presented by the Spanish Ambassador to Ireland, Dr José A. de Yturriaga, in Dublin, and the government gave Distinguished Service Medals to Paul Kellett, and posthumously to Michael Quinn. A plaque was also given by the Spanish Navy which was mounted on the bridge of the *Deirdre*. When the ship was decommissioned, the plaque was transferred to a new dining complex built at the Naval base. It had been dedicated to Leading Seaman Quinn's memory at a ceremony attended by the Minister for Defence, Michael Smith, on 19 February 1999.

That very same month, the Sikorsky and Air Corps crews were decorated for their actions in the *Dunboy* rescue. Captain Nick Gribble and winchman John McDermott received silver medals at the first award ceremony hosted by the Irish Marine Emergency Service. Appreciation awards for meritorious service were given to co-pilot Carmel Kirby, winch operator Peter Leonard and Captain Donal Scanlon and his crew of the Air Corps Dauphin. McDermott wasn't put off flying by the experience; he is now a captain flying helicopters for his own company based at Shannon.

THE SITUATION IN WHICH John McDermott found himself on the *Dunboy* should, in theory, never happen again. All winchmen are now equipped with hand-held radios which allow them to communicate with the crew above. Other technological improvements have made the job of winching crews somewhat safer, but it is still a high-risk profession — as the British Coastguard is all too aware. It lost one of its most experienced winchmen, 50-year-old William Deacon, on 19 November 1997 during the rescue of ten crew from a ship which ran aground off the Shetland Islands.

The *Green Lily* was drifting, disabled, off Bressay in mountainous seas with 15 crew on board. The Lerwick lifeboat was launched and the Stornoway helicopter was scrambled to stand by the vessel while tugs tried to reach it and take it in tow. Coxswain Herwin Clark took the Lerwick lifeboat alongside and held it there until five of the crew were saved, even though it was slammed several times against the hull by heavy seas. By then the ship was only 200 metres from shore.

The Stornoway helicopter then flew in to take off the other ten men, and the last were being taken up when winchman Deacon was washed overboard and drowned. Lerwick lifeboat coxswain, Herwin Clark, was awarded the RNLI's gold medal — the first to be given in 16 years — and his crew received bronze medals from the institution. Deacon was given the George Medal posthumously. It was the first loss in the British Coastguard's history since it was reconstructed as the Maritime and Coastguard Agency.

A SIMPLE UMBRELLA, RATHER than any sophisticated technology, saved the life of a Donegal woman in a rescue up in the north-west the year after the *Dunboy*. The Finner helicopter was called out on 15 August 1994 when Rose Solly was reported missing. Mrs Solly had been walking in the Blue Stack Mountains with her husband and a relative, and had struck ahead of the two men. As Flight Sergeant Christy Mahady recalls, she veered off on the wrong trail. The men only noticed she was missing when they reached the end of the walk and discovered she wasn't there before them. The two men climbed back up to look for her, but had to take shelter from a storm.

Little did they know that Rose Solly had broken her leg in a fall over a small precipice near Lough Belshade. The alert was raised about 1 a.m., but low cloud prevented the Dauphin from being used. At first light the Dauphin took off and flew mountain rescue volunteers up to the Blue Stacks. 'We decided to go around the far side of the mountain, just to eliminate it from the search area, though no one believed she would be over there,' co-pilot Captain Dave Swan remembers. Pilot on the mission was Commandant Donal Cotter. 'Something caught Donal's eye — an umbrella waving furiously. She was in behind a rock. We had her winched off 30 minutes later, when thick cloud was coming down.'

Christy Mahady sent Airman John Forrestal down to get her. She was in 'remarkably good spirits' for someone who had spent over 12 hours out in the elements with an injury. 'She was very hardy,' Mahady says. 'She was tipping away in the back as we flew her to Sligo Hospital.' For some years afterwards, Mrs Solly sent Christmas cards to the Air Corps squadron, and her husband wrote a letter to the then Minister for Defence and the Marine, David Andrews. 'But for their excellent efficiency on the Bluestack Mountains on the morning of 15 August, my family would have unquestionably been without a much loved wife and mother', Mr G. C. Solly wrote. It is one of only a handful of 'thank you' letters the Air Corps has received for its efforts.

10

LOST AT SEA — THE CREW OF THE *CARRICKATINE*

Cats' paws start as rough spots on the water filled with diamond-shaped ripples or capillary waves, which can catch the wind on an otherwise calm sea. With a breeze of just six knots, turbulence can begin to build — and build, as author Sebastian Junger describes so graphically in *The Perfect Storm*.

Helicopter pilots who have spent years with the British military, often flying out over the North Sea, say they have never seen anything like the waves on a bad day out in the Atlantic. And those waves are getting worse. Freak 80 and 90 footers are becoming much more common according to the scientists who study the subject. Some attribute it to the greenhouse effect, or to the impact of El Niño, and some experts quoted by Sebastian Junger blame it on a reduction in oil pollution and a drop in plankton levels. Why? Because although oil may be environmentally harmful, it can literally calm troubled waters by preventing capillaries from being generated. Plankton has the same effect because of a chemical it produces.

The Donegal fishing vessel, the *Carrickatine*, was not Junger's apocryphal *Andrea Gail*; nor was it ever going to fish so far out. However, the skipper in Junger's novel would have understood the pressures Jeremy McKinney and his crew were under when they put to sea from Killybegs, Co. Donegal, in November 1995.

The Greencastle vessel had been fishing on the Stanton banks, some 50 miles north of Malin Head, Co. Donegal, on the day it disappeared. A number of other vessels working in the area returned to port because of adverse weather conditions, and it was assumed the *Carrickatine* intended to do likewise. On board were skipper Jeremy McKinney (27) of Moville, his brother Conal McKinney, John Kelly of Ballymagroarty, Derry, his son Stephen, Terry Doherty of Greencastle and Bernard Gormley, also of Greencastle.

McKinney, one of a family of five, was acknowledged as a skilled and competent fisherman who had completed the skipper's course at the National Fishery Training Centre in Greencastle four years before. He had leased the *Carrickatine* about six months previously, and he and his brother Conal were said to have been interested in buying it at some stage. In the meantime they had to make their trips pay.

Conal McKinney (29) had worked with computers in Britain but had decided to return home to fish. He played the guitar and was a good musician. John Kelly (38), the only married man in the crew, had fished for many years and lived with his wife Josephine near Carndonagh. The couple had six children, aged from six to 18 years, and their second eldest, 16-year-old Stephen, was on board with his dad for that trip. Stephen was just starting out fishing, having finished his Junior Certificate at Carndonagh Community School.

Terry Doherty (23) was raised in North America of Irish parentage, had worked in a US bank, and the family had returned to Greencastle. A cousin of the Kellys, he had been fishing for the past few years. Bernard Gormley (18) celebrated his birthday the day the *Carrickatine* disappeared. He was the third eldest of six children and had taken up fishing two years before. William Brown, another crew member, didn't make that particular trip.

The 25-year-old 85 foot steel vessel wasn't in great condition when McKinney leased it from Denis Boyle. Originally built in Germany in 1970, it was designed for side trawling and was registered as the *Thunfisch* in Hamburg. When it was brought to Ireland in 1991, it was modified to produce a working deck area aft which was covered in a layer of concrete; towing arrangements were also altered and a new generator set was fitted, among several other changes.

The vessel lacked some basic safety equipment, and the VHF was troublesome. In his last successful radio communication on 15 November 1995, McKinney reported an engine failure and a list. He told another vessel by radio that he had had to stop the engine for a while to clean an oil filter in the gearbox. The engine had 'clutched out', he was reported to have said. When it was running again, the ship still had a list which he would need to 'sort out'. He did not specify the extent of the list, or whether it was to port or to starboard. It was the last broadcast from the *Carrickatine*.

The alarm was raised in the early hours of the following morning when the vessel failed to return to Greencastle. The previous night a north-east gale was blowing in the Irish Sea when a Howth fishing vessel, the *Scarlet Buccaneer*, got into trouble while running for port. Skipper Timmy Currid of Wellington Bridge, Co. Wexford, regularly fished the 50 foot wooden vessel owned by fellow Wexford man Kevin Downes, out of Howth. He was an experienced fisherman and he and his three crew — John White, David Kruse and Martin Murphy — knew the area well.

Shortly before 3 a.m. on 16 November the *Scarlet Buccaneer* was approaching Howth to shelter from the weather when it was caught by a strong south-westerly drift and ran aground on the seaward side of the east pier. The skipper was thrown across the wheelhouse, the engine stopped and the vessel started to break up. Wind at the time was north-north-east force seven to gale force eight, the seas were rough, and the skipper sent out a Mayday on VHF channel 16.

Patrick Kelly, who was in his lorry on Howth pier, witnessed the incident at about 3.05 a.m. and called 999. Howth lifeboat was called out by the Marine Rescue Co-ordination Centre (MRCC) and was on the scene at 3.27 a.m. but couldn't reach the vessel from the seaward side. It asked for a helicopter to be called out. At 3.29 a.m. the MRCC made contact with the duty officer at Baldonnel but was told that no suitable helicopter was available. Since the Dauphin had been moved to Finner to cover the north-west, only Alouettes with daytime capability were available at the Air Corps headquarters.

Commandant Sean Murphy was on duty at Finner when a call came through from the MRCC in Dublin. Could the Dauphin assist the vessel in difficulty just off Howth Harbour? Murphy knew that the weather conditions would involve flying around the coast and down the Irish Sea rather than directly overland, due to the poor visibility and low cloud base. It would take at least 90 minutes to get there from Donegal. He suggested that the RAF in Wales would be a quicker option. The MRCC agreed and called Swansea Coastguard at 3.36 a.m. An RAF Sea King was tasked from Valley in Wales.

The four crew who were all wearing lifejackets couldn't see the lifeboat, as the *Scarlet Buccaneer* was lying at a 60 degree angle on its port side, the seas were breaking over it and the hull was being washed up and down the side of the east pier. Several men on the pier tried to get a breeches buoy to the vessel and a line was passed out. Three of the crew got hold of it but couldn't hold on to it as the vessel was sliding up and down 10 to 15 feet with each breaking wave.

At 3.45 a.m. Martin Murphy called his father on the ship's mobile phone to say goodbye. The wheelhouse was almost flooded and the stern of the vessel had broken off when David Kruse decided to try reaching the shore. Clinging to the fire hose, he left by the aft end of the wheelhouse, but lost his grip and was dragged out to sea. Fortunately he was then washed ashore downwind of the *Scarlet Buccaneer* and was rescued shortly before 4.30 a.m.

Two of the three remaining crew also decided to try their chances at about 4.16 a.m. John White was washed ashore about the same time as David Kruse, but Timmy Currid was in a bad state when he was hauled out, and later died in hospital. Martin Murphy climbed on to the roof of the wheelhouse and clung on to the wire inlet vent and the ladder, though he was continually submerged by the waves as the vessel broke up. Just in time, at 4.45 a.m., the RAF helicopter from Valley arrived and winched him off. Within 52 minutes of grounding, the vessel had broken up.[1]

At 4 a.m. the scramble phone rang again at Finner, and once more it was the MRCC. A trawler was overdue in Greencastle. The call initiated what may have been the largest and most detailed and comprehensive sea search operation in the history of the State, involving units of the Irish Marine Emergency Service, the Naval Service, the Air Corps, the Irish Lights tender, *Granuaile*, the RNLI and local fishing vessels.

Commandant Murphy, co-pilot Captain Tom O'Connor, Sergeant Ian Downey and Airman John Forrestal were airborne at about 5 a.m. and flew by the coast around to north-west Donegal. The aircraft was 'getting hammered' by down-draughts a good distance from Slieve League, Murphy recalls. He estimated that it would ease once they were further west, outside Rathlin O'Beirne Island. They flew into a 30 to 40 knot headwind up by Arranmore and Bloody Foreland and began searching north of Malin Head to about 40 nautical miles.

'From there we flew back into Greencastle, and we searched the coastline from Magilligan Point back to Rathlin O'Beirne. The lifeboats, local vessels and cliff and coastal units were also out, but there was no sign. About three hours into the search we landed at Fanad Head for more fuel and then began another three-hour search. After six hours there was still nothing — no debris, not a trace. We refuelled at Carrickfin airport before completing another hour, this time focusing on the coast from Bloody Foreland back to Rathlin O'Beirne. After seven hours of flying we were too tired to continue, and we were relieved by another helicopter crew. The Casa fishery patrol plane was also called out to assist.'

The fact that the vessel had no emergency position-indicating radio beacon, no 'float-free' facility on its two liferafts, and that its VHF radio was reported to have been causing difficulties, did not contribute to the crew's chances of survival, the subsequent investigation said. The *Carrickatine* didn't even put out a distress signal before it disappeared. Whatever happened, it was probably 'rapid and devastating', leaving no time for the crew to reach the deck.

The search covered 51,000 square miles of ocean, extending from County Donegal to County Clare, and an underwater scan covered 1,200 square miles. The Naval patrol ship LE *Deirdre* began the underwater search on 24 November, and in early December it was relieved by the LE *Aisling*. In all, four Naval patrol ships were involved, and the only other comparable underwater search was that for the Aer Lingus Viscount which fell into the sea off the Tuskar Rock in 1968.

Apart from several fish boxes which were washed ashore on Tory Island, a gas cylinder and a pound board from the vessel's deck, nothing was found. The use of underwater search equipment in the latter stages was hampered by the fact that there were about 100 charted wrecks from both world wars in the area, according to Lieutenant Commander John Leech of the Naval Service.

In mid-December there was some speculation of a breakthrough when a possible contact was identified at a depth of 50 metres, Leech recorded. There was no wreck charted in this position. The Irish Lights tender *Granuaile* was called in with its larger, powerful and sonar-equipped, remotely operated vehicle (ROV), a superior piece of equipment to that then on board the *Aisling*. 'To everyone's disappointment, the contact was a lodestone rock of a size similar to that of the *Carrickatine*.'[2]

There was a suggestion that the vessel might have been towed under by a submarine; after all, this had happened to several vessels, mainly in the north Irish Sea, in the late 1980s. The investigation, which ruled out massive structural failure or an explosion, said it wasn't able to establish whether there were any submarines in the area at the time. However, it said that even if there were, proof of a collision would require examination of the wreck.

No bodies were ever found, which was particularly distressing for the bereaved families. The search had revived memories of a similar alert for the Donegal fishing vessel, *Árd Carna*, in May 1983. Then, British and Irish rescue helicopters were involved in the hunt for the fishing vessel, which was eventually located by an Air Corps Beechcraft, and the five fishermen on board were hailed as being 'back from the dead'.[3]

Less than a fortnight after the *Carrickatine* search first began, the Finner Dauphin crew on duty were out on a night training flight when they received a 'pan pan' message on marine radio from Malin Coast Guard to say that a fire had been spotted off Glengad Head on the Malin Peninsula by the local coast and cliff rescue team. The pilot, Captain Andy Whelan, had no reason not to take it seriously, given the source of the information. With co-pilot Tom O'Connor he carried out an extensive search extending 30 square miles north-east of Malin Head. The all-night search included landing for a refuel at the Fanad Head lighthouse helipad owned by the Commissioners of Irish Lights.

At around midnight an RAF search and rescue Wessex helicopter from Aldergrove joined in at the request of Malin Coast Guard. 'We flew for another three hours and saw nothing. But given that the *Carrickatine* was still missing, we knew that we would be requested to resume searching at first light,' Whelan recalls. He had three options: return to Finner which would involve a 90-minute round trip; land at Eglinton airfield over the border in Derry which would involve the RUC securing the helicopter, making arrangements for a bed and breakfast and also providing security for the crew; or land at Shackleton barracks at the British Army base at Ballykelly, north-east of Derry. Ballykelly was used as a forward refuelling base for the Aldergrove Wessex and would therefore be able to accommodate the helicopter.

'Sleep was our main concern, given the late hour,' Whelan remembers. He decided to request the hospitality of Her Majesty's government, and calls were put through to the appropriate diplomatic channels by Air Corps staff at Finner and by the RAF Wessex crew out on the search. Whelan estimated that Ballykelly would have had at least two hours' notice when the Dauphin landed — which it did sometime after 2.30 a.m.

The RAF staff were very welcoming and two of them refuelled the helicopter and offered the crew a cup of tea in their accommodation, which was based in an old airfield. Exhausted after a long night's flight, the Air Corps crew accepted but afterwards just wanted to get to bed. They knew they might snatch three hours' sleep at the very most. They were taken by a Land Rover to the Army side at the military base, where a Welsh regiment duty officer, who had undoubtedly been pulled out of his bed, greeted them and then went into an adjoining room to phone his superior officer.

'It was when we overheard him talking to the joint British Army/ RUC headquarters in Derry and telling them that he had "all four" and was keeping us under observation, that we realised there might be a problem,' Whelan says.

'Andy asked if he could ring Finner,' O'Connor recalls. 'We could see the lights of Donegal twinkling on the lough from the window, and yet it could have been a million miles away.'

The four crew, still in their immersion suits, were subsequently escorted to a secure communal rest room designated for members of the base's 24-hour guard. 'We giggled as we stumbled in a darkened room, trying to find four available empty bunks among beds already occupied by resting soldiers in uniform. Camouflage really works!' Whelan says. It was not the best short nap the Air Corps crew ever had, and they were glad to be airborne at first light.

'I slept very fitfully in my bunny suit for what was left of the night,' O'Connor says. 'I was woken by a bugler outside the window announcing reveille at 5 a.m. We got back into the Land Rover which, I remember, had the "confidential phone number" printed in red on the sides.'

The experience was mentioned back at Baldonnel, the crew earned the nickname 'the Ballykelly Four', and RAF colleagues had heard all about it when the pilots met up with them at a subsequent joint north-south search and rescue exercise in Belfast Lough. It was a particularly humorous talking point, given the close relationship between the Air Corps and the RAF over the years. There was an acknowledgment that the British Army had to be over-cautious, given the political situation in the North. Ballykelly was the location of one of the worst incidents in 30 years of conflict in Northern Ireland. In December 1982 the Irish National Liberation Army (INLA) bombed the Droppin' Well pub disco, killing 17 people — 11 of whom were soldiers stationed at the Shackleton base, and six local civilians. And in February 2000 there was an attempted bomb attack on the base.

The pilots heard that the duty officer at Ballykelly that night had been sent out to 'count penguins' on the Falkland Islands; and an apology came several weeks later from British Army headquarters at Lisburn.

WITH THE LOSS OF Timmy Currid off Howth, the *Carrickatine*

The wreck of the coaster, MV *Greenhaven*, which ran up on rocks at Roaninish Island off Donegal in March 1956. Ten men were winched to safety by the British Navy in the first recorded helicopter rescue in Irish waters. (PHOTO: FRANK SHOVLIN)

Preparing for a test flight on one of the first Alouette helicopters purchased for the Air Corps and delivered to Casement Aerodrome, Baldonnel, west Dublin, in November 1963. From left: Lieutenant J. P. Kelly, Jean-Pierre Couturier (instructor), Lieutenant Fergus O'Connor, Lieutenant Chris Carey and Commandant Barney MacMahon. (PHOTO: AIR CORPS AND LIEUTENANT COLONEL J. P. KELLY)

Air Corps winchman Sergeant Dick Murray on a record medical evacuation by an Alouette some 70 nautical miles west of Valentia, Co. Kerry, on 21 August 1974, when a Spanish fisherman was flown to hospital. The crew were Captain Hugh O'Donnell (pilot), Flight Sergeant Paddy Carey and Sergeant Murray. The photo was taken from an RAF Nimrod aircraft which was providing top cover. (PHOTO: RAF AND SERGEANT DICK MURRAY)

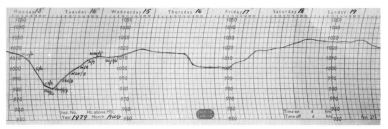

The barograph reading kept by Commodore John Kavanagh from the LE *Deirdre* on the night of the Fastnet yacht disaster showing the extent of the storm that hit the Irish south-west coast. (PHOTO: JOE O'SHAUGHNESSY, *CONNACHT TRIBUNE*)

A Dauphin helicopter about to land on the helipad of the Naval Service flagship LE *Eithne* off the Irish coast. (PHOTO: AIR CORPS AND LIEUTENANT COLONEL J. P. KELLY)

Aran Island doctor and medical officer with the Aran Island lifeboat, Dr Marion Broderick, who has worked for several decades on emergencies with the RAF, the Air Corps and the Irish Coast Guard. (PHOTO: JOE O'SHAUGHNESSY, *CONNACHT TRIBUNE*)

Joan McGinley (now O'Doherty) who spearheaded the successful west coast search and rescue helicopter campaign initiated in 1988 after the death of John Oglesby, skipper of the *Neptune*. (PHOTO: NUTAN)

RAF Flight Sergeant Mark Stephens and Air Corps Flight Sergeant Daithi Ó Cearbhalláin on the deck of the *Capitaine Pleven II* after it grounded off the north Clare coast in April 1991. The mood on board the French fish factory ship was 'calm', as was the weather, Ó Cearbhalláin recalls, and it was one of those rare occasions when there was time to take a photograph or two. (PHOTO: DAITHI Ó CEARBHALLÁIN)

An Air Corps Alouette helicopter on a training exercise. (PHOTO: MICHAEL FEWER)

The Air Corps and RAF crews involved in the *Capitaine Pleven II* rescue were decorated by the British and French authorities. The crews received the Edward and Maisie Lewis Award from the Shipwrecked Fishermen and Mariners' Benevolent Society in 1992. The RAF crew standing behind include: Flight Lieutenant Mike Baulding, captain of the Sea King, pilot officer Flight Lieutenant Richard Hooper, Warrant Officer and winch operator Peter Williams, winchman Flight Sergeant Mark Stephens and aircraft technician Ian Slater. The Air Corps kneeling in front are, from left: Captain Sean Murphy, Corporal Christy Mahady, Commandant Harvey O'Keeffe and Corporal Daithi Ó Cearbhalláin. (PHOTO: DAITHI Ó CEARBHALLÁIN)

The hull of the MV *Kilkenny* which capsized in Dublin Bay in November 1991 after a collision with the MV *Hasselwerder* close to Dublin Port. Three lives were lost. (PHOTO: PETER THURSFIELD, *THE IRISH TIMES*)

Irish Coast Guard co-pilot Carmel Kirby at Galway Airport after the near-ditch during the *Dunboy* call-out in December 1993. (PHOTO: JOE O'SHAUGHNESSY, *CONNACHT TRIBUNE*)

Inspecting the damaged rotor blade of the Irish Coast Guard Sikorsky S-61 after the near-ditch during the rescue of the crew of the *Dunboy* on 9 December 1993 off the west coast. (PHOTO: JOE O'SHAUGHNESSY, *CONNACHT TRIBUNE*)

The 'perfect' storm: the fishing vessel *Sonia Nancy* drifting helplessly in heavy seas before the rescue of its crew by an RAF Sea King helicopter in July 1998 some 190 miles off Castletownbere, Co. Cork. (PHOTO: RAF NIMROD, KINLOSS)

An RAF Sea King approaches the *Sonia Nancy* photographed by an RAF Nimrod overhead. The helicopter crew who were decorated for their efforts were: pilot Flight Lieutenant Al Potter, co-pilot Flight Lieutenant Ian Saunders, winch operator Sergeant Dave Watson and winchman Flight Sergeant Pete Joyce. (PHOTO: RAF NIMROD, KINLOSS)

The crew of the Dauphin helicopter which crashed, killing all four on board, on 2 July 1999 at Tramore, Co. Waterford, while returning from a successful rescue mission. Standing with them, less than a day before the fatal crash, are the crew of the Alouette helicopter which had provided daytime cover at Waterford up till 1 July 1999. From left: Sergeant Paddy Mooney, Corporal Niall Byrne, Captain Mick Baker, Captain Dave O'Flaherty, Lieutenant Colonel Aidan Flanagan, Brigadier General Patrick Cranfield, Lieutenant Brendan Jackman, Sergeant Des Murray, Airman Tom Gordon and Corporal Alan Regan. (PHOTO: PATRICK CUMMINS)

Maria and Davina O'Flaherty, widow and daughter of the late Captain Dave O'Flaherty who died in the Tramore Dauphin crash on 2 July 1999. (PHOTO: ALAN BETSON, *THE IRISH TIMES*)

The Air Corps rescue of seriously injured climber Emily Hackett off Ardmore, Co. Waterford, on 10 October 1999. The Alouette almost crashed during the first rescue attempt when a blanket was whipped up into the rotor blades. However, after an emergency landing the three crew — Captain Shane Bonner, Sergeant Ciaran Murphy and Corporal Aidan Thompson — decided to continue with the rescue. Ms Hackett made a full recovery. (PHOTO: FRANK POWER)

Irish Coast Guard winchman Noel Donnelly after his marathon rescue of the crew of the *Milford Eagle* on 31 January 2000. (PHOTO: LIAM BURKE, PRESS 22)

Aerial shot of the Spanish fishing vessel, the *Arosa*, after it ran up on the Skerd rocks in north Galway Bay in October 2000. Only one of the crew of 13 survived. (PHOTO: IRISH COAST GUARD AND LIAM BURKE, PRESS 22)

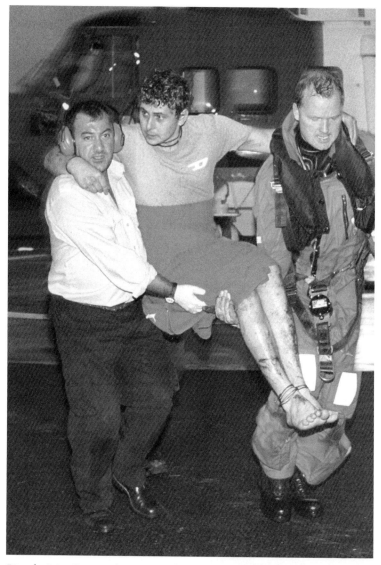

Ricardo Arias García, sole survivor of the *Arosa* which sank in north Galway Bay on 2 October 2000, being carried into University College Hospital, Galway. On the right is Irish Coast Guard winchman, Eamonn Burns. (PHOTO: JOE O'SHAUGHNESSY, *CONNACHT TRIBUNE*)

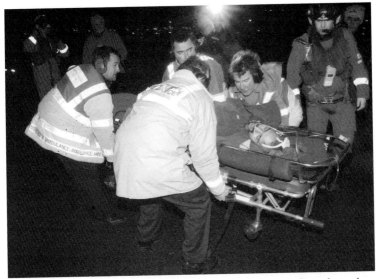

Carlos Hernandez, the seriously injured Uruguayan crewman from the tanker, *Princess Eva*, being carried into University College Hospital, Galway. His ship lost two crew in a deck accident when it hit a storm 130 miles off the Irish west coast in late January 2003. (PHOTO: JOE O'SHAUGHNESSY, *CONNACHT TRIBUNE*)

The Irish Coast Guard crew which rescued two anglers off the Isle of Man on 1 September 2003. From left: winch operator Derek Everitt, co-pilot Liam Flynn, winchman Alan Gallagher and pilot Captain Brian Brophy. (PHOTO: IRISH COAST GUARD, DUBLIN BASE)

Two happy anglers in a helicopter: Bill Hepburn (left) and John Gowan, who survived 25 hours in the water before being rescued by the Irish Coast Guard on 1 September 2003. The two, wearing lifejackets and neoprene dungarees, wisely stayed with their vessel which was kept afloat by an air lock. (PHOTO: IRISH COAST GUARD, DUBLIN BASE)

The Shannon-based Irish Coast Guard Sikorsky helicopter training with the RNLI Aran Island lifeboat and the Kilkee community inshore rescue boat in Kilkee, Co. Clare, in 2003. (PHOTO: IRISH COAST GUARD)

Captain Liam Kirwan, director of the Irish Coast Guard. (PHOTO: IRISH COAST GUARD)

Irish Coast Guard personnel with former marine minister Dr Michael Woods at Dublin Castle. From left: Eugene Clonan, Declan Hearne, Tommy Myler, John Murphy, Captain Geoff Livingstone (deputy director, Irish Coast Guard), Ben Gaughan, Martin Collins, Liam Brophy, Frank Murphy, Niall Ferns, Minister Woods, Norman Fullam, Billy Dickenson, Michael Guilfoyle (assistant secretary at the Department of the Marine), James O'Mahony, Denis Byrne, Captain Liam Kirwan (director of the Irish Coast Guard), Peter Sinnott, Eamon Torpay, Cyril Thorpe, Christy Burke, Martin Furlong, Joe Corish, Brendan Pocock, Cliff Windsor, Ger Walsh and Chris Reynolds. (PHOTO: IRISH COAST GUARD)

disappearance brought to nine the number of deaths at sea in these waters in one month in 1995. It sparked a bitter row over the ageing state of the whitefish fleet, the appalling conditions in which skippers and crew were expected to work under EU restrictions on fleet size, and the lack of 24-hour helicopter cover on the east coast. The Killybegs harbour master, Captain Philip McParlin, who resigned in protest at government inaction, said the *Carrickatine* was an accident waiting to happen. Fishing vessels were surveyed only on fleet entry and there were no requirements on stability, radio, load line or use of the latest safety equipment, he said.

The then minister for the marine, Sean Barrett, initiated several reviews — one on east coast safety and one on the state of the fishing fleet, which yielded a comprehensive report and bolstered the case at EU level for a whitefish fleet renewal package.

A funeral for the missing *Carrickatine* crew was held on the Naval flagship, LE *Eithne*, off the Donegal coast in February 1996. In County Wexford, Carmel Currid, Timmy's wife, decided to form an organisation named LOST (acronym for Loved Ones of Sea Tragedies) which would give support and information to people bereaved by tragedies at sea.

One year after the *Scarlet Buccaneer* and the *Carrickatine* incidents, a memorial Mass was celebrated in Moville, Co. Donegal, and a memorial stone to commemorate the loss of over 200 people at sea, including the six crew of the Greencastle fishing vessel, was near completion in the Greencastle Maritime Museum. That same week the Minister for Defence and the Marine, Sean Barrett, promised to implement the recommendations of a comprehensive review of air and sea search and rescue at a cost of £19 million (€24 million) over several years. A key recommendation was the provision for the first time of a medium-range helicopter for all-weather search and rescue on the east coast

The minister said he was 'determined' to provide the aircraft in Dublin by the end of next year, given that over 20 million people crossed the Irish Sea by air and sea annually, and that the volume of traffic was rising steadily. Existing Air Corps short-range cover at Baldonnel would be redeployed to the south and south-east coast. Publishing the review group's report, the minister did not rule out a role for the Air Corps in the new east coast service which would be managed by the Irish Marine Emergency Service. 'The Air Corps is currently providing an excellent 24-hour cover service from Baldonnel,' the minister said, when he stated he was 'actively considering' a number of options. A final decision would be taken 'in the context of public expenditure policy'.

The review group, which was chaired by Captain Liam Kirwan, director of the Irish Marine Emergency Service, also recommended: the provision of a fixed-wing aircraft for support including the dropping of illuminating flares, liferafts and survival equipment; the identification of helicopter landing zones on the east coast; the establishment of a marine coastal response unit at Howth Harbour to complement existing units staffed by local volunteers; the provision of vehicles for selected coastal response units; the provision of improved marine radio communications on the east coast; the initiation of a marine safety awareness initiative; and the provision by ferry owners of one ship a year to assist in search and rescue unit exercises.

The official report into the *Carrickatine* was published in November 1998 by Dr Michael Woods, Barrett's successor in a new Fianna Fáil-Progressive Democrat coalition government. It couldn't determine the cause of the vessel's disappearance but highlighted safety inadequacies on board. The publication elicited an angry reaction from legal representatives of the bereaved families. Paudge Dorrian, a Donegal solicitor, said his clients would be taking the matter to Europe.

Paudge Dorrian claimed the report had not dealt adequately with certain matters, specifically the modification of the 25-year-old boat. He criticised the delay in publishing the report and said that department officials had stated they would only meet relatives without their solicitor being present. The relatives had been told in September 1997 — almost two years after the disappearance of the vessel — that they would receive copies in a matter of weeks, but this had not occurred.

Questions remained in relation to the original construction of the boat, Dorrian said. The relatives still did not know the details of the modification; who carried it out; or who owned the boat at the time. There was 'only one marine surveyor for the entire west coast of Ireland', he added.

The minister, Dr Woods, defended the findings and said they were based on all available evidence. Should any further evidence relating to the loss of the Donegal vessel come to light, this would be followed up immediately, a spokesman for the minister said. 'The book' was 'not closed'.

The report on the death of Timmy Currid and the loss of the

Scarlet Buccaneer was published by the minister in March 1999. By then the east coast had its 24-hour search and rescue helicopter base, provided on contract to the government. In a comment on the report, Carmel Currid said the night-flying helicopter and the initiation of her organisation, LOST, represented a fitting tribute to her husband Timmy — 'a wonderful husband and father'.[4]

11

SEA CAVES AND CLARE CLIFFS

David Courtney is one of the most experienced air-sea rescue pilots in these waters. A County Offaly man, he joined the military after finishing school. Ten of his 14 years flying helicopters with both the Air Corps and the Irish Coast Guard were spent on night rescue. He served as chief pilot at the coast guard's Shannon rescue base for three and a half years.

When he left the Irish Coast Guard in 2001 to work with a commercial airline he had 4,500 flying hours, of which 750 were on the Air Corps Dauphin and 1,700 on the Sikorsky S-61. He had clocked up over 300 emergency rescue missions and had won a number of awards.

Many of the rescues which hit newspaper headlines were, he says, 'technically straightforward'. Others which might have made a paragraph at most were often far more challenging. 'If a vessel is stuck on a rock and it isn't moving, the greatest difficulty is in getting there, as in navigating at night, in bad weather, and close to cliffs and rocks in bays and inlets.'

But some of the most fantastic missions have involved small vessels without power, he explains. 'Bigger vessels don't move so much vertically, and so the size and length of the craft makes all the difference, and whether it is under power or not. If it is a fishing vessel with no engine, for instance, you're hovering over a moving deck — violently pitching and rolling in some cases — with no back-up and no light. It could also be very cluttered with deck gear. The winchman has to get down and get casualties back up to you, and you're watching the time and your fuel gauge. You might do six call-outs like that every winter. Nothing would be said afterwards, and you couldn't go looking for a clap on the back, but you'd know yourself that you had really pushed it.'

All aircrew are trained to deal with varying situations, but rescue pilots might be woken at three in the morning in very bad weather and given only snatches of information, unlike commercial pilots who have had their rest and know their routes.

'Every rescue pilot is trained to behave in a robotic manner, but he or she is also human and the safety of the crew is always the overriding factor,' Courtney observes. 'Crew sometimes have to turn a rescue down. It is the hardest thing to do, but sometimes it is the best option and the only option.'

One of the more memorable missions which occurred during Courtney's time at Shannon involved a call-out at night to Inis Óirr on 17 March 1996. 'Yes, St Patrick's Day — and Mother's Day!' Mairéad ní Fhlatharta remembers. She was in labour with her fourth child. There was no doctor on the island, but two nurses travelled with her and her husband Mícheál in the helicopter.

It takes only minutes for the Sikorsky to cross Galway Bay to University College Hospital, Galway. Mrs ní Fhlatharta had a feeling she wouldn't have minutes on this occasion. Sorcha was born in the back of the helicopter as Captain Al Lockey flew east. 'We were about halfway across,' winchman Peter Leonard remembers. 'One of the nurses was a midwife, so we just helped out as best we could with towels and things.'

The Ó Flathartas didn't want any publicity at the time, but it was the first mission of its type in Irish air-sea rescue — and the only one at time of writing — though there have been several close calls on flights from the islands. Several decades previously there had been a birth on the ferry *Naomh Eanna*. 'The helicopter crew were great, and it was a sort of a distraction,' Mrs ní Fhlatharta says. 'I really didn't have time to think about the pain. And I was back home in two days.'

WHEN DIRECT HELICOPTER ASSISTANCE proves to be impossible, co-operation between rescue agencies can often make all the difference. One such collaboration occurred on the north Clare coastline on the evening of 23 September 1997. Clare ambulance control put an emergency call through to the Irish Marine Emergency Service (IMES) in Dublin to report that a young boy had fallen from a path near the bottom of the Cliffs of Moher and was seriously injured.

Peter Fitzgerald (13) was with his brother and a friend on a steep patch at around 8 p.m. when he slipped and fell about 40 feet on to a ledge. The Irish Coast Guard unit in Doolin, Co. Clare, was tasked, and an ambulance crew from Ennistymon was also called out, but it became clear that the young boy couldn't be carried back up the path. An Australian nurse who was in the locality managed to join the crew on the ledge and had assessed his condition.

Mattie Shannon of Doolin, who has been involved in many rescues on the treacherous Clare coastline, sought help from the Aran Island lifeboat and the Shannon-based Sikorsky. Noel Donnelly and Pete Leonard, members of the helicopter winch crew, both remember the conditions. 'The wind was easterly, there was no ambient light, the pilots couldn't see the cliffs and they couldn't see the lights of the coast and cliff rescue unit either,' Donnelly recalls. 'The boy was about 500 foot down on the north side of the cliffs, towards Doolin. We couldn't get close to where he was, and neither could the lifeboat. We made about four or five attempts, but you can't fly in close to the cliffs unless you can see them and unless the wind and visibility is suitable.' It wasn't suitable that night.

The easterly wind would have made winching down to the location almost impossible, according to Peter Leonard, even if they had been able to approach it. The turbulence was too great. The coast and cliff unit went down and hauled the young boy in a stretcher up the 500 feet. It emerged that he had sustained a broken collar bone and other injuries; he was fortunate to make it, as a 27-year-old man had fallen to his death from the cliffs the previous Sunday.

'We landed in a field and we airlifted him to hospital when they arrived. It was all done in pitch dark and was quite an achievement by the coast and cliff team,' Leonard says. Observers pointed out that the lights provided by the lifeboat and the helicopter were crucial to the success of the rescue.

Mattie Shannon and his team subsequently received a bronze medal at the State's inaugural marine merit awards, when a bronze medal was also given to Michael O'Regan, area officer of the IMES coastal unit and his crew at Goleen, Co. Cork, for the very difficult recovery of the body of a German tourist, Dietrich Versl, from the cliffs at Mizen Head lighthouse in January 1994.

Those inaugural marine awards were named after a diver, the late Michael Heffernan, who was also involved in a co-operative effort on the north Mayo coast in October 1997. On 25 October the emergency services received a report that four people were overdue at Belderrig pier. Will Ernest von Below (53), a retired German businessman living at Belderrig, near Ballycastle, had taken Tony Murphy, his wife Carmel and their 11-year-old daughter Emma out for a trip in his new 16 foot currach. They went exploring the sea caves at the foot of the cliffs at Horse Island when the vessel got into trouble. The party managed to get into one of the larger caves but were trapped when the currach was overturned by a freak wave.

The alert was raised at 5.30 p.m. and the Ballyglass lifeboat, the Irish Coast Guard unit in Killala and local fishing boats began to search the coastline, with the gardaí and local people searching the shore. The Irish Marine Emergency Service tasked the Shannon Sikorsky helicopter, flown by David Courtney, co-pilot Captain Gerry Tompkins, winch operator John Manning and winchman Noel Donnelly, at 5.40 p.m. The missing group was located by a local man, Padraig McAvock, and the search party at around 7 p.m. that night when they heard shouts and whistles coming from a tideline-level cave on Horse Island.

Superintendent Tony McNamara, then second coxswain of the Ballyglass lifeboat, co-ordinated the rescue and called for the Garda Sub-Aqua unit to assist. There was no mobile phone coverage in the area, so he spent the night working on a hand-held VHF radio — and using a phone in a private house about three-quarters of a mile away.

The Sikorsky helicopter, which was helping to illuminate the cave, was tasked to fly to Dublin and pick up the Garda divers. But there wasn't time. The IMES asked some members of the Gráinne Uaile diving club, based in Ballina, to help. Club members Michael Heffernan, Josie Barrett, Michael Kelly and Noreen Roleston travelled to Belderrig.

As the Sikorsky was *en route* to Dublin, Sean McHale, coxswain of the Irish Coast Guard Killala unit's rescue boat, took the Gráinne Uaile divers to the cave mouth and stood by as two of them, Josie Barrett and Michael Heffernan, swam in with a light line. It was an incredibly brave attempt; there was a real fear that the people trapped inside would perish when the tide filled the cave. The divers hoped to swim in and haul the people out before that happened.

Sea conditions in the cave were horrendous, with waves up to 12 feet high. The divers became separated; one of them, Josie Barrett, managed to swim back out and was picked up by Michael Kelly and Sean McHale in a state of complete exhaustion and taken to the Ballyglass lifeboat. It was assumed that the second diver, Michael Heffernan, had reached the missing party.

The helicopter returned with the Garda divers, and McNamara's main concern was to ensure that it had a suitable landing site. 'If it had had to fly to Ballyglass to land, we could have lost another hour, so we needed to be sure it landed as close to the pier as possible,' he recalls.

The superintendent contacted members of the Belmullet and Crossmolina fire brigades and asked them to illuminate a field close to the pier. He even got its precise latitude and longitude co-ordinates for the helicopter pilot from a local fishing boat, which used its Global Positioning System (GPS) receiver to work out a true bearing from seawards to the landing site.

'The Garda divers were fully kitted out when they landed, and that saved valuable minutes also,' McNamara says. Sean McHale and crewman Martin Kavanagh picked them up and took them to the cave mouth. Several times the boat tried to re-enter without success. They devised a strategy whereby Martin Kavanagh would advise McHale of the following waves; McHale would then turn the boat and meet the waves head on. On the third attempt the inflatable was tossed up and thrown on to boulders at the back of the cave, 15 feet above the water level.

They found the Murphy family huddled together in a crevice just over three feet above sea level. When they were all thrown out of the currach, Tony Murphy found the body of Mr von Below beside him in the water at the back of the cave. McHale, Kavanagh and Garda divers Ciaran Doyle, David Mulhall and Sean O'Connell eventually managed to haul the inflatable to the water's edge and relaunch it, while attempting to reassure the Murphy family. They discovered that the dinghy's engine was damaged beyond repair. Sadly, Garda Doyle found the body of Michael Heffernan close by.

At this point it was going to be extremely difficult, if not impossible, for anyone to get out, and they knew there was no immediate possibility of rescue from outside. The gardaí talked to McNamara on VHF. There was only one option. Ciaran Doyle, one of the diving team, decided to swim out with a 250 metre line attached to the inflatable and connect it to one of the fishing vessels at the cave mouth. As he set off, Mulhall and O'Connell helped the Irish Coast Guard crew to get the survivors on board the damaged inflatable.

'This was a mammoth undertaking, literally a life or death attempt,' McNamara says of Garda Doyle's action. 'It was a swim of almost 1,000 metres or 3,000 feet in darkness and appalling sea conditions, while pulling a heavy climbing rope behind him. It required extraordinary courage, strength and the ability to brave it.'

Skipper Patrick O'Donnell of the *Blath Ban*, who was watching out, spotted Doyle emerging. He recovered him, and the line, from the water. Working with Martin O'Donnell of the *Sinéad*, the *Bláth*

Bán skipper took the tow line and hauled the IMES inflatable out to safety. 'Atrocious' was how Doyle described the situation inside the cave afterwards.[1] 'It was pitch dark in there. It was very helpful that the family were wearing lifejackets. They were very intelligent and had got into a crevice at the back of the cave about a metre above the water line and away from the extreme danger. The child was not panicking and the parents had done a good job in consoling her. We reassured them all and brought them out,' he said.

Word spread then among the waiting rescuers that Gráinne Uaile diver, Michael Heffernan, had not survived; nor had the German, Will Ernest von Below. Michael had been hurled against rocks in the cave in his attempt to reach the Murphys. The survivors were taken on board the lifeboat, skippered by deputy second coxswain Gerard Reilly, and transferred to Belderrig pier. The Sikorsky then flew them to hospital in Sligo, refuelled at Finner camp and returned to base at Shannon.

Patrick O'Donnell of the *Bláth Bán* helped to recover the bodies of Michael Heffernan and Will von Below. The O'Donnells, the Garda and local divers, the IMES Coastal Unit crew, the Ballyglass lifeboat coxswain and the Sikorsky crew all received awards at the subsequent IMES awards ceremony, with the highest accolade being accorded the diver who had died.

At the State's first Marine Meritorious and Long Service Awards in Dublin Castle on 26 February 1999, hosted by marine minister, Dr Michael Woods, the gold medal was given posthumously to Michael Heffernan. Silver medals were presented to Garda Ciaran Doyle, Sean McHale and Josie Barrett of the Gráinne Uaile diving club. Bronze medals were given to Garda David Mulhall and Garda Sean O'Connell, and to Martin Kavanagh, McHale's assistant on the IMES coastal unit.

In a letter to *The Irish Times* the following month, Captain Liam Kirwan, IMES director, said that at least 21 services and 100 people had participated in the rescue, including helicopter crew, divers, climbers, the Gárda Siochána, the fire service, the ambulance service, medical staff, refuelling centres and lighthouse staff, Aer Rianta and air-traffic control. Malin coast radio station had provided the communications, which was co-ordinated by Dublin. 'In addition, an enormous contribution was made by local fishermen and members of the local community.'[2]

There was considerable shock over Michael Heffernan's fate, as he was the first civilian volunteer to die in a marine rescue. There was a consensus among rescue personnel that his death should be commemorated, and so six crew from various rescue agencies teamed up to try and break the round Ireland speed record which had stood for 11 years. They hoped to raise funds during the circuit for the Heffernan family, and funds would also be given to the Irish Brain Research Foundation and the Irish Council for the Blind.

Garth Henry and Gerard Bradley from the RNLI lifeboat station at Portrush, Co. Antrim, David Courtney and John Manning from the Shannon helicopter base, Sean McCarry, a paramedic with the British Coastguard in Northern Ireland, and Robert Stirling from the Northern Ireland health service volunteered to crew a ten metre Delta rigid inflatable for the attempt, and Charlie Cavanagh and Michael Doherty from the Greencastle IMES team said they would monitor progress and issue press information. The crew achieved their target, setting out on 28 August 1998 from Portrush and returning in 21 hours and 54 minutes — beating the previous record by 11 hours and 40 minutes.

The Garda divers received Scott medals for bravery, and they attended a ceremony almost three years later at Lacken pier in north Mayo, when Annamarie Heffernan, Michael's widow, unveiled a solid bronze statue in his memory by sculptor Nick Hughes of Foxford. Mrs Heffernan was accompanied by her two children, Leigh Anne (6) and Michelle (3). She was pregnant at the time her husband died. Kevin Myers of *The Irish Times*, who was asked to perform the unveiling because of an article he had written on the Belderrig rescue in 'An Irishman's Diary', said the request represented for him the greatest honour.

Also among those who attended was David Courtney. The Sikorsky crew had been given a merit award by the Minister for the Marine for their role in the rescue. The citation said it 'was probably one of the longest missions performed by a single crew and most of it at night' and 'great skill and determination' had been shown.

'I think we played more of a supportive role at what was a very public and emotional event,' Courtney said afterwards. 'We have had other 13-hour rescues which wouldn't have had such a profile. There were an awful lot of people involved on that day who made much more of a contribution than we did.'

Superintendent McNamara disagrees. 'The Sikorsky's task that night in diverting to Dublin to collect the Garda divers and to land them prepared, with all their diving equipment on, in a boggy field at night on a hill overlooking Belderrig pier was vital. David

Courtney and his crew did an excellent job as time was of the essence.'

It was a fabulous co-operative effort, McNamara emphasises, though those who worked with him on the night know that this teamwork was in no small part due to him. 'What had made things even more difficult was the fact that it was the night that winter time came in and the clocks changed, and we were having to work on estimated times of arrival with limited communications. All the services, professional and voluntary, worked so well. It was a tragedy to have lost two lives, but it could have been more. Josie Barrett came within seconds of losing his life; only that he was strong enough to pull himself out, he wouldn't be here. And he was heartbroken to hear about Michael, who was such an excellent diver and so competent in every situation. Josie was sure he would have made it back out, but the sea was just that bad.'

THE SEA CLIFFS OF north Clare — the impressive, treacherous Cliffs of Moher — marked the location for another memorable co-operative rescue in April 2000, but this time the Shannon helicopter was tested to its limits. On 15 April a team of 32 climbers had volunteered to abseil down the cliffs for charity. They had all completed one successful descent by lunchtime, and six climbers had just finished their second abseil in the afternoon when there was a rock fall. Two climbers, who were being assisted off their ropes by a safety officer at the time, were buried. The safety officer was seriously injured.

The Irish Coast Guard unit at Doolin and the Shannon helicopter were called out, and the Coast Guard escorted the four uninjured but very shocked climbers at the foot of the cliff back up, in the knowledge that another rock fall could be imminent. The helicopter crew were on a 'limsar' or limited search and rescue; pilot Captain Simon Cottrell, co-pilot David Courtney and Eamonn Burns, winch operators, were working as a trio because their winchman was ill.

'We got the call that some people had been injured in a rock fall, and we thought we should go out and stand by for some air-ambulance work — presuming that the cliff rescue team would get the injured parties to the top of the cliff. We had no idea of the situation we were to come across. We had no idea we were going to do any winching,' Eamonn Burns says.

'We did a fly-over and it was just the next bay north of the O'Brien stack. On the flare camera I could see there was at least one person under a big boulder, and another person being attended by the cliff rescue team who seemed to be unconscious. Given the height of the cliff and the seriousness of the situation, we decided to do a recce without disturbing the cliff rescue team. There was a bit of down-draught, we were just off the shore, and we came back out. The captain asked me if I'd be able to carry it out, and I said that of course I would, but it would require a bit of co-ordination. In other words, the captain of the aircraft would have to become the winch operator from his seat, communicating with me, while the co-pilot flew the helicopter in as close as he could and held it in hover,' Burns says.

'It was the strangest feeling. I was there working in the back of the aircraft doing a search and rescue on my own which had never been done by this unit before. I organised my medical kit, my stretchers, got myself to the door and checked my safety equipment, while Dave hovered the aircraft as close as he could. There was water underneath me, which was safe enough. I got to the door and looked out. The captain saw me in his rear view mirror and I gave him the nod. He winched me down as far as the water, and when I disconnected he winched in and the aircraft flew away. I was on my own then, and I carried the stretcher over to the casualty and worked with the coast and cliff rescue team. We stabilised him somewhat, and then when he was ready we got the cliff rescue team to carry him into a position where we could take him up. We were winched out of the water up to the door of the aircraft, and this was the really difficult part,' he continues.

'Normally it takes the strength of my legs and the winch operator's hands to pull the stretcher away from the door and allow access into the aircraft. With a heavily laden stretcher, I had to manoeuvre myself into a position, and I didn't have a winch operator to grab the top of the stretcher, pull backwards and pull me into the aircraft. It was one of those situations where I found myself in the back of the cab with the casualty but have no idea how I got there. But anyway it was done. I disconnected and we flew the casualty to Galway, giving treatment all the way. He was critically ill. He was in a coma for a couple of days and in intensive care for at least two weeks. The last I heard he had recovered. When I was treating him, he had severe head injuries and multiple fractures throughout his body — and open fractures in most of the bones of his left leg,' Burns says.

'The location was lucky, in that if it had been anywhere else further up on the cliffs, and had the tide been in, it would have been very different. We have a 300 foot cable, 295 foot of it usable, but we still might not have been able to fly in. It was just about on the border for us. It was an extremely dangerous manoeuvre on everyone's part because there were only three of us.'

Doolin Coast Guard dug out the two bodies and brought them up the 600 foot cliff face with the assistance of Kilkee Coast Guard unit, because the walking pathway was too narrow. They did so in the knowledge that there could have been a further rock fall. Both Doolin Coast Guard, led by Mattie Shannon, and the Shannon rescue helicopter received marine bronze medals in November 2002 for the 'exceptional skill, commitment and determination' that both rescue units showed in recovering the seriously injured climber from the foot of the cliffs.

12

CALM PATCH, BUILDING . . . OH JESUS CHRIST!

Sometimes a simple sequence of events can make such a difference to people's lives. So it was in the case of the *Sonia Nancy* off the Cork coast in early 1998.

Flight Lieutenant Alan (Al) Potter was on duty at the Royal Air Force (RAF) base in Chivenor, north Devon, when the alert was raised at 7.25 a.m. on 4 January 1998. He was already awake, wandering about the operations room, having been startled by a loud bang. The room had been hit by lightning, and all the phones, faxes and meteorological computers were dead. 'The only functional piece of equipment was a mobile phone, and so I called RAF Kinloss to tell them what had happened and to pass on that number,' he remembers.

RAF Kinloss was delighted to hear from him, as it had received a report of a vessel in difficulties some 190 miles south of Castletownbere in west Cork. Would the Sea King be available to take a look? It was still dark, but Potter had a fair idea of the conditions. It had been a wild, wild night and the wind was still gusting to 80 knots. He would talk to his crew, two of whom had also been awoken by the lightning flash and had just joined him in the operations room.

Potter, originally from Worcester in the British midlands, had been flying search and rescue for six years and had been almost three years as deputy flight commander at Chivenor. He joined the RAF in 1985 and had always intended to be a pilot but wasn't accepted at first. He worked as a sergeant on the Nimrod reconnaissance aircraft for three years, got his commission in 1989 and was accepted for pilot training. Initially he wanted to fly fast jets, but towards the end of his training he realised that his heart was in helicopters — and search and rescue.

'At this point, around breakfast time on 4 January, we weren't dealing with an absolute emergency,' Potter remembers. 'So we began to plan. I asked the crew first if they were willing to risk their lives on this one — was it achievable? To a man they all said yes. I've never been in a situation where one person has said no, but you'd have to respect it if that happened, particularly if it was a winchman. We may be in the military but you can't order people out. You'd be a poor captain if you did, because in search and rescue the aircraft and crew come first.'

The crew calculated that the best option would be to fly to Cork to refuel, rather than heading directly to the Scilly Isles. Winds were just too strong for that. At this point RAF Kinloss said the vessel definitely required assistance: could the helicopter crew go? Kinloss would request diplomatic clearance for the Sea King to travel through Irish airspace.

Information was scant enough, but there were ten people on board a fishing vessel which had been under tow. The tow had broken and the powerless vessel was in danger of sinking in high winds and heavy seas. The computers were still down, so Potter put a call through to the Meteorological Office at Bracknell to ascertain weather conditions *en route*. 'I've never met the guy we talked to, but he was absolutely spot on with his information,' Potter says. 'Without that level of precision the plan wouldn't have worked. I've never experienced such accuracy since!'

The crew — pilot Flight Lieutenant Potter, co-pilot Flight Lieutenant Ian Saunders, winch operator Sergeant Dave Watson and winchman Flight Sergeant Pete Joyce — were airborne by 8.15 a.m. It took a few minutes to gauge it because the gusts of 75 to 80 knots would only drop off to 45 knots for a minute or two; 45 knots is the maximum speed at which we can start the rotors, Potter points out.

As the helicopter coasted out over north Devon, the crew looked at each other. The sea was like nothing they had ever witnessed. 'It was totally white, like drifting snow.' The helicopter was being buffeted about and the vibration in the back of the aircraft was so bad that the two winch crew began to feel sick and had to move up forward and stand behind the pilots.

Normally flying time from Chivenor to Cork would be about an hour and a half. It took the Sea King three hours and 55 minutes. At one point the captain noted that the helicopter was moving forward at a speed of 22 knots, and 58 knots to the right. 'We had to look out the right-hand window to see where we were going because visibility was just so bad.'

The rotors were still running as the Sea King refuelled in Cork

Airport. Within 20 minutes the crew were airborne again. The pilots took in a top-up or pressure refuel at Bantry in west Cork. 'When we turned crosswind, it took us another two and a half hours to get to the boat,' Potter says. 'So altogether, by the time we had reached the *Sonia Nancy*, we had been travelling for almost seven hours.'

One of two RAF Nimrod reconnaissance aircraft called up that day was already over the vessel, which was being tossed about helplessly in mountainous seas. 'The sea state was about 12, but what made it quite difficult was the massive swell,' Saunders remembers. The conditions were so bad that the *Sonia Nancy* had lost all its liferafts on deck, and almost all its safety gear, bar one lifejacket among a crew of ten.

The Nimrods had dropped liferafts, but one of the pilots reported that the *Sonia Nancy* skipper had a 'virtual mutiny' on his hands. The crew wouldn't abandon the ship, fearing for their chances in liferafts in the horrendous swell. 'Each liferaft lasted about 20 minutes, before it was ripped away from the vessel,' Potter says.

A 30,000 ton French container ship in the area, the *Fort Descaix*, had picked up the radio messages from Falmouth Coastguard, which was co-ordinating the rescue, and had also tried to help, but conditions were just too bad. The crew of the *Sonia Nancy* were terrified that they might be crushed by the container vessel's hull if it attempted to come alongside. 'We actually thought at first that the ship was the fishing vessel, which gives you an idea of the sea state,' Potter says.

There was no sign of the *Mapescal*, the vessel which had been towing the *Sonia Nancy* some hours before. 'Conditions were so dreadful that I think they lost each other pretty quickly. It was probably searching for the *Sonia Nancy* and could have been within half a mile, but the waves were just too big to see it.' Ironically, three of the Nimrod dinghies were found a few days later: one off Brest in France, one near Chivenor, and one near south Wales.

Potter remembers that the fishing vessel was lying broadside to the swell, the aft deck was partially flooded and most of the crew were in the wheelhouse. Fortunately the Irish radio operator on board, David Fox, had both English and Spanish, and had already been talking to the Nimrod. The helicopter crew made contact with him. 'I asked Pete if he was happy to go out on the winch, given the conditions, and he said yes. However, given the flooding aft, we would have to lower him on to the bow and it was pretty cluttered,' Potter says.

There was some further communication with the vessel and then, in a scene which Potter will never forget as long as he lives, two of the crew emerged from the wheelhouse roped to each other and crawled towards the bow. 'The guy in front was holding an angle grinder — how he hung on to it in those conditions, I don't know — and they cut the stays and the mast toppled. It was enough to clear the deck for us.'

The helicopter released the hi-line and fortunately managed to convince the crew that it shouldn't be tied to the deck or it would pull the helicopter down. 'They started to pull Pete in. He was wearing his backpack dinghy for emergencies, which is very heavy. Next thing, a wave hit the boat, it pitched up and the guard rail caught Pete in the back. He was stunned, and went completely limp, like a rag doll on the wire. We began to winch him in as fast as we could and took him on board. He had recovered a bit by that time and was quite happy to go out again, but I said I wasn't going to let him.'

With an eye on the fuel and the clock, co-pilot Ian Saunders briefed radio officer David Fox on how to get two crewmen into two strops and hook them on for transfer up to the Sea King. 'We sent the strops down and the guy sat in them like bosuns' chairs, so Fox had to explain that they must put them under their shoulders. The RAF Nimrod put a stopwatch on the winching operation, and it was all done in nine minutes, Fox being one of the last two to go up,' says Potter. 'We had calculated we had enough fuel to stay 45 minutes at the scene, allowing for our return. At that stage we had been 44 minutes and 30 seconds there.'

Potter says it was one of the hardest decks he has ever had to work over. 'Helicopter pilots are taught to use very small inputs in search and rescue. With this situation we were forced to apply big corrections. At one point I remember looking forward and seeing this enormous wave coming at us, which was three-quarters of the way up our cockpit window. We were sure it would sink the boat, and wondered if it wouldn't catch us.'

Saunders also remembers that roller. 'Al was busy flying, the winch operator was busy watching the ship and I was watching the swell and talking to the winch operator. I was calling out "Calm patch, building, building . . . Oh Jesus Christ!" The wave had come up as far as the helicopter, and it was at that point that Al looked up. It just missed us, but it has to be about the biggest wave I ever saw.'

A second RAF Sea King had been dispatched from Valley, flown by Squadron Leader Alan Coy. Potter and Saunders saw its lights as they moved off, discarding the hi-line as they did so. The second helicopter turned back as it saw them approaching. 'The survivors were all fine, apart from a few scrapes and bruises, and Pete Joyce was bruised and sore, but he was all right. Our original plan was to fly to the Scilly Isles, refuel and return to Chivenor, but it was getting dark at this stage and there were lots of thunderstorms around us. Whereas our speed had been limited to 90 knots at 1,000 feet on the way out, the ground speed on the way home with that wind behind us was over 200 knots. It was like being on a jetliner!' Potter says.

'The Scillys flashed up and we still had fuel. It was the first time most of us had rescued people who weren't injured, so there wasn't any need to go straight to hospital. We were told to fly to Culdrose in south Cornwall and to pick up a "press pack". I thought that was some sort of information we had to deliver for somebody. As we shut down at Culdrose and opened the front door, we were faced with a mass of flash bulbs going off. We had no idea there had been that much interest, but found out afterwards that the rescue had been on Sky News broadcasts every hour throughout the day. I have to say this wasn't great for our families — we'd prefer them not to know too much about what we are involved in on a day like that!

'We really couldn't talk to the press immediately as we all had one thing in mind. We had been nine hours and 55 minutes in the air from the time we took off to our landing at Culdrose, and yet no one could understand why we all just wanted to go to the toilet!'

Several days later at Cork Airport, radio officer David Fox was given a hero's welcome when he was greeted by his family, including his mother Breda and his brothers and sisters living in Farranree. Also there to welcome him was the owner of the *Sonia Nancy*, Patrick Sheehy from Baltimore in west Cork. He had leased the vessel to the Spanish operators and confirmed that the vessel was on the seabed with no hope of salvage.

Breda Fox was delighted and said she had been out of her mind with worry. She said her son's heroism didn't surprise her, however, as she knew just how 'strong and calm' he could be in a crisis. 'I am just pleased to have him back home,' she said.[1]

Interviewed by reporters, Fox said the ordeal hadn't put him off a life at sea, and he had been working on Spanish vessels for the past 12 years. He intended to go back working as soon as he could, knowing of course that with the sinking of the *Sonia Nancy* he was currently out of a job.

It was a 'close thing', he said, and he had 'looked death in the eye'. He knew that the helicopter had reached its maximum range and that they had a very limited time span in which to execute the rescue. 'If it had taken five minutes longer, they would have had to leave the area because of fuel shortages and we would have been finished,' he told *The Irish Times*.[2] In fact, as Potter had noted, there was a short 30-second margin.

The radio officer was critical of the training of the Spanish crewmen. 'I had to tell them exactly how to use the helicopter winches as they were not trained in any methods of rescue . . . and with time running out fast it was a case of someone having to take control.' He pointed out that Irish and British fishermen had to undergo compulsory safety training, but these rules didn't appear to apply to the Spanish crews. 'There must be some type of uniform training for those working on board trawlers, irrespective of the place of registration or the nationality,' he said.

All four RAF helicopter crew were given awards for their efforts. The captain, Flight Lieutenant Potter, and winchman Flight Sergeant Joyce received the Air Force Cross, while the co-pilot Flight Lieutenant Ian Saunders and winch operator Sergeant Dave Watson were given the Queen's Commendation for Bravery. 'Usually the whole crew don't get an award — it is normally the captain and the winchman — even though it is always a team effort,' Potter notes. 'So this was particularly good, and we saw it as something for the whole search and rescue squadron.'

The Spanish Ambassador to Britain, Señor Alberto Aza, also thanked all the RAF members involved — the four crew on the Chivenor-based helicopter, the reserve helicopter flown by Squadron Leader Coy from Valley, and the Nimrods flown by Flight Lieutenants Kev Hughes and John Meston. 'This is another example of the solidarity and collaboration between the United Kingdom and the Kingdom of Spain, and further proof of the close ties which unite our two countries', the ambassador wrote in a letter to the British Defence Secretary, George Robertson. The Defence Secretary, in his response, noted that the search and rescue flight and reconnaissance aircraft crews based around Britain 'do a selfless job often in desperate conditions to save lives. They do not seek thanks, but thanks are richly deserved.'

13

TRAMORE, CO. WATERFORD, 2 JULY 1999

Maria O'Flaherty spoke to her husband Dave on the phone at around 7 p.m. on 1 July 1999. It had been a long day. Early that morning he had left their home in Lucan, Co. Dublin, and driven to Baldonnel aerodrome for duty as pilot with the Air Corps Dauphin rescue helicopter. Maria, a bank official, had left the house for work shortly after him.

Dave had taken his overnight bag with him, but this wasn't going to be any ordinary 24-hour shift. As detachment commander, O'Flaherty (30) was due to fly that morning to Waterford Airport with co-pilot Captain Mick Baker (28), winch operator Sergeant Paddy Mooney (34) and winchman Corporal Niall Byrne (25). There they would meet the Minister for the Marine, Dr Michael Woods, and their immediate boss, the General Officer Commanding (GOC) the Air Corps, Brigadier General Patrick Cranfield. The 'top brass' would be down for the formal opening of the south-east region's first 24-hour air-sea search and rescue base.

The opening was the fulfilment of a commitment by the minister who had already done much to develop the marine search and rescue network during his term in office. The previous couple of years had seen some terrible accidents, mainly involving fishermen at sea, including the disappearance of six crew on the *Carrickatine* off Donegal, the death of Timmy Currid, the County Wexford fisherman, off Howth pier in November 1995, and the death of three crew from the 40 foot *Jenalisa* which sank off Dunmore East, Co. Waterford, in February 1996.

The 1996 review of east coast search and rescue recommended that a medium-load helicopter be based in Dublin. Given the concentration of air and ferry routes on the Irish Sea, and the growth in marine leisure, the decision was not before its time. The Royal National Lifeboat Institution (RNLI) had been providing a vital voluntary rescue service, and the west coast now had medium-range cover after Joan McGinley's very successful campaign.

Yet there were still gaps and a continued dependence on the Royal Air Force and British Coastguard for the north-west, east and south-east coasts. This was largely attributable to the inadequate equipping of the Air Corps and the failure to deliver on key recommendations in the government's own Doherty report of 1990. A separate consultancy review of the defence force in 1998 recommended the purchase of four medium-lift helicopters for the Air Corps, and advised against 'privatising' search and rescue.

None of these recommendations had been acted upon by the time Dr Woods took office, but the Irish Marine Emergency Service — which was to become the Irish Coast Guard — had been expanded, and the contracted medium-lift helicopter at Shannon was working well. 'All I want to do is to provide a service,' the minister told *The Irish Times* on 29 June 1998, when a Bond Sikorsky S-61 helicopter landed for photographs at Howth in his Dublin constituency. Bond, which was already running the medium-lift service for the Irish Marine Emergency Service at Shannon, had been a successful bidder for a new contract for the east coast.

The Air Corps, already running the north-west base at Finner camp in County Donegal, was to be deployed also in the south-east, according to the east coast review. The location made a lot of sense. Bordered by Tuskar Rock and the Saltee Islands, the south-east has long been regarded as one of the most treacherous sections of the 2,700-mile coastline and has been a graveyard for many ships over the centuries. A fishing accident in 1958 accelerated the development of Dunmore East Harbour in the neighbouring county of Waterford. Three Donegal fishermen had lost their lives when their vessel, the *Mary Buchan*, attempted to steam up the River Barrow to Waterford to take shelter from a storm. The vessel, which had been ring-netting for fish, was swamped by heavy seas just outside Dunmore East. A few years later, in yet another accident costing three lives, a local vessel, the *Mary Joseph*, was lost off Hook Head.

Rosslare, Co. Wexford, is a busy ferry port with daily passenger services to and from Wales and France. It is one of several ports and harbours in the area also overflown by substantial air traffic. One of the State's few air accidents, the crash of the Aer Lingus plane with 57 passengers and four crew on board *en route* from Cork to London, had occurred off the south-east lighthouse base of Tuskar Rock on 24 March 1968.

There had been some more recent tragedies in the area. In 1993 two teenage fishermen who were cousins — Pat Tobin and P. J. Rossiter from Ring, Co. Waterford — were lost off Helvick Head. In December 1995 two County Waterford cousins, Edmond Fitzgerald and Paul Dunne, were lost off Ardmore when their 21 foot fibreglass boat was smashed up on rocks near by. On 18 February 1995 Dunmore East was the scene of a tragic canoeing accident: two young people were swept out to sea in rapidly deteriorating weather conditions. Michael Davies, who lost his 14-year-old son Ros in the incident, was to campaign for statutory safety regulations in adventure centres.

Initially a daytime service involving an Alouette III aircraft was set up at Waterford from July 1998. A commitment was given to provide a full 24-hour service from 1 July 1999. The Air Corps was going through cyclical staffing difficulties and had a shortage of trained crews. Some of the most senior Dauphin pilots had left or were leaving to take up flying positions with commercial airlines, at a time of low morale within the Defence Forces generally.

And so the big day came, and the minister had his press release prepared; the military were not going to steal his thunder when this was a key stage in his plans. It was also a very significant day for Waterford Regional Airport. The Air Corps was under pressure. With medium-range search and rescue now contracted out on both the west and east coasts, some of the pilots and crew felt that the defence wing had been sidelined. If 24-hour deployment did not go ahead successfully on the promised date, there was a belief that it could reflect badly on them.

For all those reasons, there would be press coverage and plenty of photographs — and there were — photographs on the airfield, photographs by the helicopter, photographs of Captain Dave O'Flaherty and co-pilot Captain Baker in the cockpit. The following day some of those shots of smiling airmen stared hauntingly out of the pages of several newspapers. By then the crew of Rescue 111 had been called out on the base's first night mission. A 15 foot boat was lost off Dungarvan, Co. Waterford, in very bad visibility.

En route back after the boat had been located, the helicopter was enveloped in thick fog and tried two unsuccessful approach landings at the airport. Running short of fuel, the pilots then tried for the beach at Tramore, four kilometres away. At about 40 minutes after midnight on 2 July the helicopter crashed into a sand dune, was engulfed in flames, and all four airmen lost their lives. The first details were given on the early radio news bulletins, when those newspaper images from the previous day's ceremony were already tragically out of date.

Maria O'Flaherty, before going to bed that night, called her husband's mobile phone at about 10.40 p.m. The call was diverted. Often during his time on search and rescue duty at Finner camp, she would leave a message and ask him to phone her back — just to be sure he was safe. She didn't this time, but it did not bother her too much. The south-east did not seem quite as harsh or hazardous as the Atlantic off Donegal.

She was awoken in the early hours of the following morning. Two Air Corps officers were at the door. Dave's helicopter was missing. At about 6.30 a.m. a telephone call confirmed the worst. Her husband and his three colleagues were dead.

CAPTAIN O'FLAHERTY AND HIS crew had flown the aircraft three times on 1 July before the call-out later that night. The first flight was from Baldonnel with the four aircrew and three technicians on board. They arrived at Waterford at 10.02 a.m. and were airborne again at 11.41 a.m. when they did a publicity flight for RTÉ radio.

That afternoon, when the formalities were over, O'Flaherty called a briefing for the flying and technical crew. The mood was positive and a 'can do' spirit prevailed, according to one of the senior technicians present. However, he also remembers the detachment commander warning that the Irish Marine Emergency Service might try to 'catch them out', especially on call-out times.[1]

One of the technicians asked about ground support during night operations; the airport closed at 4 p.m. Monday to Friday. O'Flaherty told the technician that the control tower and airport lighting would be his responsibility, as he had learned just that day there would be no air-traffic control support 'after hours'. Final agreement on the payment for this was still the subject of negotiations between Waterford Airport and the Department of Defence.

There wasn't much time to think it through, but clearly this was far from satisfactory. Waterford was not a military airport; the commercial facilities were under the control of a separate management structure. Although the technician had some experience of ground/air communications when the north-west search and rescue base had been relocated temporarily from Finner camp to Carrickfin airport in Donegal, he had never been in Waterford Airport before. Nor had he any formal training in control tower, air-traffic control or meteorological procedures.

Shortly after lunch the crew carried out a training exercise south-east of Tramore. This involved some simulated emergencies over the sea. Landing just under an hour later, the aircraft took off after just four minutes to check out the route to the nearest hospital. Afterwards, O'Flaherty recorded that the weather had been 'fairly poor'. He also noted that the aircraft was 'serviceable', denoted by the letter 'S'. One of the technicians back at base had seen rolling sea fog off Tramore and asked the commander if they had been caught up in it. Conditions were grand, he was told.

The aircraft was refuelled, washed down and towed into the hangar. The aircrew and technical staff left the airport about 4.15 p.m. and headed for their temporary accommodation at the fishing harbour of Dunmore East. They stopped to shop for food *en route*. They would stay in three houses — one for the two pilots, one for the two winch crew and one for the three technicians. It wasn't quite what they were used to when on duty in Finner camp, where the military base had its own dedicated accommodation for search and rescue crews.

They had cooked and eaten when Dave O'Flaherty's phone rang just after the 9 p.m. television news. 'Hello there. MRCC Dublin here,' said the voice from the Marine Rescue Co-ordination Centre's Dublin headquarters at the Department of the Marine and Natural Resources in Leeson Lane. 'We have a job for your shiny new Dauphin.'

'Very good, O'Flaherty replied. 'I had a feeling you'd call us out on our first night.'

At first the information was pretty scant. A 15 foot boat, colour yellow, with an outboard motor and no radio on board had not returned to its port at Dungarvan. The RNLI inshore lifeboat at Helvick had been tasked by the MRCC, but the lifeboat had no radar and so initially it was unable to locate the casualty. The helicopter would have to assist it with its on-board equipment.

The aircrew were in a minibus and *en route* from Dunmore East to Waterford Airport when O'Flaherty's phone rang again. It was just 9.09 p.m. and the MRCC was going to stand the crew down. It had established mobile phone communication with the small boat and the lifeboat would locate it fairly shortly. The weather was poor with 'very bad visibility'.

'Ah, just a sec, just one moment, just one moment,' the MRCC officer continued. There was a short conversation with his colleague back in Dublin. 'Will I stand him down?' he was heard to ask.

'The helicopter?' his colleague responded. 'Well if they want to go for a bit of practice, things like that, we don't mind.'

'Are we tasking him or are we not?'

'Are they at the airport?'

'We're on the road out to the airport,' O'Flaherty said.

Another short exchange in Dublin: 'The lifeboat hasn't launched yet you know . . . Let them go, let them go.'

The helicopter, Rescue 111, was airborne from Waterford by 9.41 p.m. There was one more exchange with the MRCC in Dublin and a reference to the poor visibility. It emerged that a young child was on board the boat and was very seasick, and the skipper was heading into the weather to try and keep the craft stable. After that, the helicopter was in communication with the MRCC through several coastal radio stations at Rosslare and Minehead, and with Waterford Airport's tower. As far as the pilot was aware, the tower was manned by one of his technicians. He did not know that the airport manager, a formerly qualified aerodrome flight information services officer, had arrived in the tower at 9.55 p.m.

At 10.19 p.m. the helicopter informed the airport tower that the lifeboat had the casualties under tow. Ballycotton lifeboat was also making its way to the area close to the entrance to Dungarvan Harbour. The helicopter would not return yet, however. Because the lifeboat's navigation system wasn't working properly, it needed help from the aircraft to guide it back safely in the fog to Helvick pier, some two miles away. The commander agreed to stay 500 feet overhead and descend if the lifeboat got into any trouble. He was in 'a lot of cloud, a lot of fog,' he said.

For the next hour the helicopter was in constant touch with the lifeboat, checking on its compass course to ensure that it was heading for the harbour. At 10.41 p.m. O'Flaherty called the tower to check on the weather.

'Er, negative. She's staying the same,' the tower replied. 'From what I can see from here out to the lights, QNH [the altimeter setting to obtain elevation when on the ground] is still one zero one four and the wind is down to about seven knots at two . . . two twenty at seven knots. Over.'

'Copied,' the helicopter replied, acknowledging the message.

Nine minutes later the airport tower called the helicopter again. 'Just to inform you, weather deteriorating slightly here. Just to let you know. Over.'

'Roger, copied that . . . Can you see the lights of Tramore at all?'

'Negative. We can just about see the runway which is a distance of 300 metres from the tower. Over.'

'Roger, copied that, er listening out,' O'Flaherty replied. The pilot called Rosslare radio and asked for permission to route towards Waterford Airport because conditions were deteriorating. 'We'd like to get in before they close. Over.'

Thirty seconds later, at 10.54 p.m., Rosslare radio informed the pilot that the helicopter was released. 'Thank you for your help and your co-operation. Over.'

Helvick lifeboat also called the helicopter to express their thanks, but there would be some further communication with Rosslare radio within 15 minutes. At 11.11 p.m. the helicopter said it was approaching Waterford Airport and would be 'letting down'. It would be refuelling once landed and would be contactable by land link.

'Rescue 111, Rosslare radio, that's all copied. Thank you very much indeed. We'll update you on the situation there at Dungarvan on the land line.'

The helicopter did not land. At 11.14 p.m. it told the airport tower it had overshot the approach and was going to go around again. The tower couldn't see it, and it couldn't see the runway's landing lights. A technician was out on the ramp, but visibility was down to about 500 metres.

At 11.15 the tower asked if they had seen the runway lights. 'Negative,' was the pilot's response.

'Roger that. We'll try to assist with a higher beam light maybe from the tower. I don't know how bright it's going to be. Over.'

'If the lights are full up, that's the best you can do.'

At 11.18 p.m. the helicopter tried another approach and overshot the runway once again. 'We're going to go around for a coastal approach,' the pilot said.

'Roger that,' the tower replied at 11.27 p.m. 'We couldn't see you coming in again, but could hear you going away just as I called you. Over.'

A minute later, at 11.28 p.m., the tower suggested changing the approach lights. The helicopter agreed. At 11.33 p.m. the pilot asked if the weather was improving.

'Negative on the weather improving here,' the tower responded. 'QNH same, wind two ten at eight. Over.'

'That's copied,' the pilot said. 'We're just in a left. We're descending here now in the bay and we are just going to do a coastal approach in to Tramore. We may land in Tramore.'

'Roger that, tower.'

At 11.34 p.m. the tower called again. 'Just for your information, we have been on to Bal tower [Baldonnel tower] and the weather there is fine to get in there if all need be. Over.'

'Roger. Don't have the juice,' the pilot said. A minute later, at 11.35 p.m., there was a recorded conversation between the helicopter and Rosslare radio, updating it on the fact that it had overshot Waterford Airport due to the weather and could not get in. 'We're doing an approach to Tramore Bay this time and if we can get down we're going to land in the bay area somewhere. Over.'

'Rescue 111, Rosslare radio, that's all copied. Keep us updated please.'

'Roger. Will do,' the helicopter replied.

This was the last recorded exchange with the aircrew. At 11.38 p.m. Rosslare radio called and got no response. At 11.39 p.m. there was a short noise from the helicopter recorded on the Waterford Airport tower tape. There was no voice.

At 11.55 p.m. the Waterford airport manager and the MRCC informed Tramore gardaí that the helicopter was overdue. A Mayday relay was broadcast by the MRCC on Rosslare radio at five minutes past midnight. At 13 minutes past midnight the local coastal search team attached to the Irish Marine Emergency Service was called out.

THE 'RED ORANGE GLOW' in the dunes led them to the wreckage. Visibility was down to between 10 and 15 feet when the local shore rescue team attached to the Irish Marine Emergency Service came across it at about 1 a.m. The fog was so bad that some of the team had become disorientated on the beach, and one man had trouble using his mobile phone.

The wreckage was still burning, and parts of the helicopter were strewn everywhere. The official report by the Air Accident Investigation Unit gives a detailed, clinical account of the scene:

The aircraft, on a heading of 130 degrees magnetic (122 degrees true), impacted with a sand dune. The crest of the dune was approximately 14 metres (45 feet) above sea level.

The main rotor blades contacted the dune approximately one metre below the crest. The initial impact displaced approximately one ton of wet sand creating a sizeable crater in the face of the dune. Remnants of the fuel tanks were found in the crater. A small amount of other wreckage, including the Emergency Locator Transmitter (ELT) and the main rotor blade leading edge strips, was scattered immediately outside the crater. The signal from the ELT was transmitting and was subsequently disabled at the site by removal of its battery.

More wreckage, including the hoist hook (sheared from the cable), aircraft battery and pilot's door, was strewn between the initial impact and the crest of the hill. The majority of the wreckage (including engines and main gearbox, hoist, shells of line replacement units) was strewn down the back slope of the dune, and scattered in the valley at the base of the back slope.

Many components from the aircraft tail were found in the valley . . . The bodies of the four crew members were found amongst the wreckage on the back slope, near the crest.[2]

The fire on initial impact had been intense. Some of the aluminium items had melted, indicating exposure to temperatures of over 600 degrees Centigrade. Most of the aircraft had been consumed by the blaze and the entire site was 'littered with charred debris'. Only the tail section which had broken clear escaped the furnace. The accident was 'not survivable', the report said.

The investigation commissioned a computer simulation of the final moments from Cranfield Impact Centre Ltd, a campus company based at Cranfield University in Britain. In all cases the model showed that the aircraft pitched nose over tail, having hit the dune and slid up it, when it became airborne again. The tail then came under the cabin, and the helicopter slid tail-first down the back slope where most of the wreckage was recovered. Estimated speed was nearer the 'lower end of speed range', at 50 knots rather than 125 knots.

The rescue services arriving on the scene in the early hours of 2 July were not immediately aware of the highly hazardous nature of the area. There was the potential for a magnesium fire, the prevalence of burned or burning man-made fibres and burned synthetic fluids, and the aircraft's pyrotechnics and compressed gas cylinders which had been exposed to very high temperatures. At about 2.30 a.m. an Air Corps senior technician arrived at the site and advised the IMES rescue teams, the gardaí and fire brigade of the risks. The shoreline was enveloped in dense fog. The Sikorsky S-61 in Dublin was called in to assist and landed at Waterford at 3.42 a.m.

'They were very nice lads, very professional and very helpful.' The RTÉ cameraman and his colleague, who had taken a trip in the helicopter only the day before, were clearly shocked to be back so soon. Within a matter of hours they would be filming the men's coffins as they were taken from the crash site, known locally as the 'rabbit burrows', to Waterford Regional Hospital.

Tributes flowed in. The President, Mary McAleese, the Taoiseach, Bertie Ahern, and individual members of the Government and Opposition parties expressed their sympathies with the families. The Minister for Defence, Michael Smith, and the marine minister, Dr Michael Woods, travelled to Tramore and both were said to be visibly distressed by what they witnessed. 'The aircraft itself had flown in about 450 missions, saving hundreds of people,' the defence minister said. 'On this occasion, we're faced with the loss of four great men who went out in high-risk conditions to save people . . . They were within minutes of getting down safely when they hit the dune.'

The Department of Public Enterprise initiated the official investigation by the Air Accident Investigation Unit, while the Defence Forces initiated a separate military court of inquiry. An annual air show in Galway involving participation by the Air Corps was cancelled as a mark of respect. Similarly the planned opening the following day of a new Irish Marine Emergency Service station house at Killala, Co. Mayo, and a presentation of long service medals to volunteers in Counties Galway, Mayo and Donegal was postponed. The RAF offered its sympathies and said it would provide cover on the south-east coast if required.

A Sea King flew to Baldonnel the day after the crash, stayed for five days and attended several of the funerals. 'It was a very moving time for everybody, and it could so easily have been us,' said Squadron Leader Al Potter of the RAF, who had won an award for his role as flight lieutenant, along with his colleagues, in the RAF Sea King rescue of nine Spanish and one Irish crewman from the Cork-owned fishing vessel, the *Sonia Nancy*, off the Irish south-west coast the year before. A unit of the Royal Navy's Fleet Air Arm joined members of the Air Corps in Kinsale, Co. Cork, at a multi-denominational service where prayers were said for the dead men and their families. The Royal Navy personnel who had been in Cork as part of an exercise to demonstrate their rescue capabilities said they had been deeply moved by the tragedy.

One of the many tributes came from the Aran Islands general

practitioner, Dr Marion Broderick, long-serving voluntary medical officer to the RNLI's Galway Bay lifeboat. She had depended for many years on medical evacuations provided by Air Corps, RAF and Irish Marine Emergency Service helicopter crews: 'All islanders depend for our safety and security on the selfless dedication of the men and women who provide our rescue services,' she said. 'Because of their total professionalism, commitment and willingness to respond in impossible weather conditions, it has never been safer to live on islands. Without them, our survival as island communities would be threatened. My heart goes out today to the wives, children, parents and families of these brave men, who have made the ultimate sacrifice to serve others. Our thoughts and prayers are with them in their grief.'

The four had been among the most skilled in the force, the newspapers reported on 3 July. Captain O'Flaherty from Tullamore, Co. Offaly, had been with the Defence Forces for over 11 years and had started flying with the Dauphin in 1997. In February 1998 he was part of a rescue mission which had tested the aircraft's limits. The Shannon-based Sikorsky had answered a call-out on 25 February to lift an injured fisherman from a Norwegian vessel, the *Leinebjorn*, off the Mayo coast in 50 knot winds with a 20 foot swell. The Sikorsky's controls registered a gearbox failure and it was forced to abandon the mission, leaving its winchman behind on the vessel.

The Air Corps Dauphin at Finner camp in the north-west was called out as relief, as was an RAF Nimrod, two RAF Sea King helicopters and another S-61 Sikorsky from Shannon. Commandant Sean Murphy, Captain O'Flaherty, Corporal John O'Rourke and Airman Jim O'Neill escorted the Sikorsky back to Blacksod on the Mayo coast and refuelled. The Dauphin then flew out to the fishing vessel to airlift the injured man and the Sikorsky winchman from the heaving deck. The Air Corps crew flew the Norwegian to hospital in Sligo. Their helicopter had barely touched down when the crew received a request to pick up a sick man on Inishbiggle Island, Co. Mayo, and bring him to Castlebar — which they did.

Dave O'Flaherty loved sport, particularly basketball, and was keenly interested in the hurling fortunes of his native Offaly where his mother, Lily, still lived. He had some 2,910 flying hours under his belt on all types of aircraft, some 808 of those on the Dauphin. He had recorded 138 hours of flying at night. One of his more unusual taskings had been in response to a call about a man 'in a sheet' who was stuck up on a tower in the old Dublin lead mines.

The crew thought it was somebody who might need gentle persuasion to descend. In fact it was a young man on his stag night who couldn't remember how he had got up there!

Dave O'Flaherty was listed on the Defence Forces press release as being 'married with no children'; however, he had known that his wife Maria was expecting their first child.

Captain Mick Baker from Enniscorthy, Co. Wexford, had joined the Air Corps in 1988, and with 2,327 flying hours was one of the most experienced for his age in the service. One of a family of five, he loved literature, played several musical instruments including the trumpet, and was a keen artist at school. On finishing school, he won a place in law at Trinity College, Dublin, but he had always been keen on a military career and on flying. He was accepted for a place at the Royal Military Academy in Sandhurst, but opted for the Air Corps. When he was offered a cadetship, he asked his father's advice. 'I knew he wanted to fly,' his father, Tony, said, and so he advised him to accept. His long-term aim was to return to study law at Trinity and fly part-time for a civilian airline, as he had a civilian aviation licence.

Described as a 'fanatical' GAA fan, Mick was particularly keen on hurling. He played Gaelic sports with several clubs including the local Rapparees and the Air Corps. He was one of the Air Corps team to take part in the Round Ireland yacht race — twice. At the time of his transfer to Waterford he was living in Castleknock, Co. Dublin, with his long-time girlfriend, Siobhán Dunne.

Sergeant Paddy Mooney from Stamullen, Co. Meath, was married to Monica. She was also from Stamullen and they had known each other since primary school. The couple had three children, Aisling (10), Conor (8) and Mark (2). Paddy had joined the Air Corps in 1981 and was very popular at Baldonnel aerodrome. He had also served in the Curragh with the Army Rangers. He had saved many lives during his time as a winch operator, clocking up 3,500 flying hours as rear crew, and he was also a search and rescue aircrew rating examiner. He had served overseas with the Defence Forces as part of the 55th Infantry Battalion in the Lebanon.

Among his many missions had been a rescue in very heavy seas on 4 November 1996 off the Mayo coast. 'One of those white knuckle jobs' was how Commandant Sean Murphy, the pilot on duty at Finner camp that night, described it afterwards. A fisherman with head injuries on board the *Benchourn*, some 75 miles west of Blacksod, required airlifting to hospital in a westerly storm blowing to 75 knots. The helicopter's only communications link, other than the fishing vessel, was the semi-submersible exploration rig, Sedco 711, which was then drilling on the Corrib gas field off Achill.

Two hi-lines snapped during the rescue effort and the crew only had one left when Airman Jim O'Neill reached the deck and found himself sliding around in the horrible conditions. 'I think he assessed the casualty as being able to come off using a strop,' Murphy said afterwards. 'Once hooked up, I remember yanking both off with the finesse of a catapult.'

The helicopter, flown by Murphy and Captain Kevin Daunt, had just enough fuel on board to get to Sligo Airport, but not to the hospital. On approaching Sligo, Paddy Mooney, the winch operator, noticed that the hi-line was trailing. The crew were convinced they had discarded it successfully at the scene. It had become snagged in the undercarriage; fortunately it hadn't done any damage. That same night, after the Dauphin returned to Finner, the hangar roof blew off. The helicopter had to be filled with sandbags to weight it down, and the unit was transferred to Carrickfin airport some 70 miles away in west Donegal the following day.

The youngest of the four who died at Tramore was Corporal Niall Byrne, single and living at home with his parents and family in Killiney, Co. Dublin. A keen canoeist, he paddled in several Liffey Descent races. He had originally joined the Army and had served overseas in the Lebanon with the 78th and 82nd Infantry Battalions. On returning from his second Lebanon trip, he forfeited some leave to start training with the Air Corps which he saw as a great opportunity, according to his father, Vincent.

Niall only transferred to the search and rescue service the previous December and had 175 flying hours as rear crew. The posting to the new south-east 24-hour rescue base was his first major operational assignment. One of five children, he was engaged to Teresa Bolton and they had made a deposit on a house. They planned to get married in September 2000 on Niall's parents' wedding anniversary.

In the initial aftermath there were strenuous denials that equipment failure or fuel shortages had contributed to the crash. The Minister for Defence pointed to the extent of the inferno and said this indicated there was still significant fuel on board. He would not have been party at that point to the final recorded communications with the aircrew in relation to a possible return to Baldonnel.

At this point, it was known that the helicopter had remained with the Helvick inshore lifeboat and the small angling boat to ensure a safe passage home, but there was much speculation as to the actual cause of the crash. Both ministers admitted to reporters that they were baffled as to how a Dauphin with high-tech navigational and search and rescue equipment should fail to land despite the fog.

On Sunday, 4 July, over 1,000 people turned out to pay their final respects to the relatives of Captain Dave O'Flaherty. The chief mourners were his mother Lily, his wife Maria, his brother Dermot and his sister Valerie. Earlier that day President McAleese had led the mourners at a funeral Mass at Our Lady of Loreto Church, Casement aerodrome, Baldonnel. The Air Corps chaplain, who had presided at the wedding ceremony for the young O'Flaherty couple just two years before, recalled how he had enjoyed a meal with them only a fortnight ago.

The chaplain, Father Brendan Madden, described the young pilot as 'a man who lived life to the full and who only ever wanted to do good for others'. His life had 'not ended, only changed', the chaplain said. During the offertory procession gifts brought to the altar included a photo taken with his mother when he got his wings commission, his football boots, a bottle of red wine, personal gifts which the couple had exchanged and photos of their dog.

14

MISSING MAN FORMATION

'It was a beautiful summer afternoon in a grassy place between the mountains and the sea. Quiet and peaceful. A lovely lazy day for watching jet planes draw white vapour trails across the sky. . . . To the uninitiated, it was a fitting and impressive tribute to a dead hero. But the ranks of airmen saying goodbye to their brave colleague could read the other message in the sky. Because the three Marchetti aircraft were flying overhead in Missing Man formation.'[1]

So wrote Miriam Lord of the *Irish Independent*, witnessing the heartbreaking funeral of the youngest member of the aircrew, Corporal Niall Byrne, in Killiney, Dublin. Chief mourners were his father Vincent, his mother Anna, his sister Aoife then aged 15, his brothers Eunan, Donal and Ronan, and his fiancée Teresa who clutched his red Manchester United football jersey and his sky blue Dubs jersey before they were presented as offertory gifts.

Three priests officiated at the Mass, including the corporal's cousin, Father Frank Monks, who was to have conducted Niall and Teresa's wedding the following year. He described Corporal Byrne as a unique individual, 'loving, serious, single-minded, generous to a fault, religious in his own quiet but convinced way'. He thanked President McAleese for attending the service and pointed out that the corporal had been very pleased to meet her when she had visited the Irish UN Battalion with which he was serving in the Lebanon in December 1997. Among the pilots flying in the Marchetti formation had been Squadron Leader Al Potter of the RAF.

Even as Tony and Mary Baker were preparing to lay their son Michael to rest, burglars broke into their home at Carley's Bridge near Enniscorthy, Co. Wexford. They had travelled to the removal at Baldonnel air base and escorted his coffin to the parish church in Enniscorthy. Celebrant Fr Tony O'Connell told a packed congregation at the removal service that 'nobody wants to die but if there is a good place to die, what wonderful company to die in, at the service of people in difficulty.'

Tragically, it was the Bakers' second time to bury one of their children. John Charles Baker died from a heart condition in September 1980 at the age of three and a half. His father had always been grateful for the honesty of the consultant at the time, when he found he could do no more to save him. Michael was buried beside his baby brother in Enniscorthy Cemetery.

Pupils from St Patrick's National School in Stamullen, Co. Meath, formed a guard of honour as the remains of Sergeant Paddy Mooney arrived on 6 July. Members of St Patrick's GAA club and the Balscadden Blues soccer club were also out in force. Over 800 mourners paid their respects to the popular sergeant's widow, Monica, and their three children. Among the pall-bearers were his brother, Private Joe (Dave) Mooney of the 5th Battalion, Gormanstown, and his cousin and close friend, Private P. J. Cudden of the 29th Battalion.

The Minister for Defence, Michael Smith, pledged a speedy replacement for the Dauphin helicopter which had been destroyed in the crash. He told *The Irish Times* on 4 July that another Dauphin would be stationed in Waterford to resume the 24-hour service which the government and the Defence Forces had been committed to providing for some time. Whatever had to be done to make this come about 'would be done', he told the newspaper, whether it involved purchasing a new helicopter for the area or relocating a Dauphin from elsewhere in the State. It was likely to be 'some months' before the night-time service could be restored. This was not simply a matter of replacing the lost aircraft; there was also a considerable investment to be made in producing crews of the calibre and experience of the men killed the previous week.

In the event, 'some months' would be 'some years'. It would be a full three and a half years before a night-time service would resume at Waterford — only to be postponed in its final stages, owing to a threat of the airport's closure. In the interim there would be a tortuous, highly political and unseemly row over a replacement aircraft, following the publication of a damning report on the causes of the crash. It was the worst accident in Air Corps history. Seven months after Tramore, Maria O'Flaherty gave birth prematurely to her first child; the baby girl was named Davina.

THE FIRST RUMBLINGS FROM a defence wing which had been devastated by the events of 2 July 1999 came two months after the crash and a full year before the publication of the official investigation. In late August of that year a former Air Corps commanding officer warned that the Dauphin helicopter fleet must be replaced by aircraft more suitable to the Atlantic margin. Lieutenant Colonel Ken Byrne, who had recently retired after over 30 years of service, said that larger, long-range craft would allow for a safer 'manoeuvre envelope' on search and rescue missions. The Dauphin helicopters had served well for 13 years but did not have the endurance or reserve power for the job required, he told *The Irish Times* in an interview published on 30 August 1999.

'The Dauphin has been described as state-of-the art, but this only relates to its navigational system and avionics. There is no point in having this equipment if the craft itself is short on endurance and reserve power. This coastline was known for its particularly varied and localised weather conditions, which made certain weather changes very difficult to forecast. When such conditions change without warning, a bigger helicopter with greater endurance and bigger power/weight ratio increases the safety margin.'

Asking the aircrew to do the job it was required to do with current equipment was similar to 'asking the Royal National Lifeboat Institution to run with 40 foot power boats' in very hostile coastal conditions. Ken Byrne also said that excessive bureaucracy and lack of a clear direction from the government on the Air Corps future was having a detrimental effect on morale at a time when there were already many more opportunities elsewhere in the aviation industry.

On 19 October 1999 the Minister for Defence told the Dáil that it was unlikely that Dauphins would be used again by the Defence Forces in search and rescue work. However, he said that replacement would be in the context of the impending White Paper on Defence and the implementation of the Price Waterhouse consultancy review of the Air Corps.

Several weeks later the minister was made aware that this response was not good enough. Commandant Aidan Flanagan, Officer Commanding the Air Corps helicopter wing, chose an award ceremony hosted by the Irish Security Industries Association in Dublin on 17 November to air his views. Flanagan was receiving the association's premier award on behalf of the Air Corps helicopter wing, when he said it was particularly welcome at a time of low morale in his unit. This was a product of 'inaction' which had resulted in many staff leaving to take up posts in the private sector.

Referring to the loss of the four airmen, Flanagan appealed to the minister and his department to make a decision on replacing the Dauphin fleet with medium-lift helicopters suitable for night-time maritime search and rescue. He said he was asking the minister 'before it was too late'. He urged that a decision be taken before the publication of the White Paper on Defence.

The GOC Air Corps, Brigadier General Cranfield, was called in by the minister to explain Flanagan's outburst and ordered to furnish a report. The background to the comments would not emerge until almost four years later, when a memo obtained from the Defence Forces, reported in *The Irish Times*, indicated that the GOC had opposed the initiation of the 24-hour base at Waterford on 1 July, two weeks beforehand. In the memo, dated 15 June 1999, he said the necessary accommodation and hangarage at Waterford Regional Airport wasn't ready. His opinion was challenged in a response two days later, on 17 June 1999, from Flanagan, who said he 'entirely disagreed' with the GOC's strategy.[2]

Three former Air Corps pilots expressed further concerns about the Dauphin fleet in an interview with *The Irish Times* published on 24 January 2000. They expressed alarm at the government's apparent lack of any sense of urgency about the helicopter's capabilities. The aircraft had inadequate fuel capacity, was not suitable for night search and rescue off this coast, and was not equipped for a return to base in poor visibility, they said.

It was a fast and very effective VIP aircraft, used in search and rescue by the US Coastguard, Icelandic Coastguard, Norwegian civil search and rescue, and the Hong Kong police; but as one former pilot noted, asking it to undertake search and rescue offshore in the Atlantic was like 'trying to use a Porsche to pull a plough'.

The official investigation was still continuing in January 2000 when Tony Baker, father of co-pilot Mick Baker, made public his discovery of a letter from his son warning of 'serious implications' for the Dauphin fleet. This was due to the difficulty in finding spare parts for its obsolete electronic flight information system and the radar equipment. 'Prior to Christmas last year, we had a situation where one Dauphin was missing vital avionic components for a period of time due to this issue', Baker had written in a letter dated 25 January 1999. 'This scenario is likely to happen on a more frequent basis due to the increasing difficulty with maintaining support as the aircraft get older and our spares situation depletes further', he warned. 'It will also mean that our ability to support a Waterford detachment this year will be very difficult.' Mr Baker found a second letter from his son detailing fuel contamination during an exercise in the north-west.

Although the official investigation was being carried out by the Air Accident Investigation Unit (AAIU), it was agreed that a suitably qualified officer of the Air Corps would lead it. Lieutenant Colonel Thomas Moloney was appointed to head a team that included an inspector from the British air accident investigation branch and a military psychologist from the Swedish armed forces.

The final report released in late September 2000 was wide-ranging and uncompromising. It identified as contributory factors the lack of ground support for the airmen and inadequate safety and training procedures. Those responsible for 'serious deficiencies' in support included Air Corps senior management, the Department of Defence, and the Department of Marine and Natural Resources.

The 103-page document identified two active causes, six contributory causes and nine systemic causes for the crash. While the main cause was identified as collision with a sand dune after an unsuccessful approach to Tramore beach at night in extremely poor visibility, the report was unable to determine the reason for the prior descent of the aircraft. However, its author said that the weather in Tramore Bay was so bad that a successful landing would have been 'virtually impossible'.

Weather information emerged as a key factor. The Marine Rescue Co-ordination Centre operations memorandum for helicopter tasking, dated 21 September 1993, states that a flight forecast from the meteorological service at Shannon must be obtained for 'All Missions'. The report said this was not done for this search and rescue mission.[3]

The first weather-related communication between the Dauphin crew and the control tower at Waterford was one hour into the mission on the night of 1 July, it said. In fact it was to emerge in the separate military inquiry that attempts had been made to obtain weather information. The airport fire officer, who had arrived at the terminal at the same time as the Air Corps detachment, turned on the SITA machine to request an updated weather forecast by telex. When he returned to the building after carrying out a runway inspection, he met with Captain Baker who greeted him with 'Weather?' However, the SITA machine hadn't printed an updated forecast. The fire officer tried again but without success.[4]

The personnel in communication with the Dauphin from the tower were the technicians supporting the maintenance of the helicopter on detachment, the AAIU report noted. 'They had no training in or experience of meteorological observation. These technicians unwittingly found themselves, through no fault of their own, placed in a situation outside their training and experience.'

The report said it was possible that had an experienced air-traffic officer with a meteorological observation rating been on duty, he or she could have communicated details of poor visibility and low cloud base earlier, and even recommended diverting from Waterford.

There was much more. The report found that pressures on the detachment commander to accept the rescue mission had been 'very high' and crew fatigue was a contributory factor. The call-out had come at the end of a day of official and press duties to mark the initiation on 1 July of the 24-hour south-east base and the crew had been on duty for sixteen and a half hours at the time of the accident. It highlighted the failure to conclude an agreement between the Department of Defence and the airport on providing after-hours air-traffic control staff, and it criticised the lack of an accommodation block or catering for the crew at Waterford itself.

The report referred to a lack of sufficient 'in-theatre' training for the four crew, and other infrastructural deficiencies. The crew had decided to take 600 kilograms of fuel, believing that the mission would be carried out quite quickly given its proximity to the airport. A further 200 kilograms of fuel could have been taken, and this would prove crucial later when the captain reported that he did not have enough 'juice' to divert to Baldonnel or Shannon for a safer landing.

The report made 25 safety recommendations, including the urgent establishment by the Department of Defence of a fully resourced air safety office in the Air Corps which should be headed by a flying officer of lieutenant colonel rank. The 1998 Price Waterhouse consultancy review of the Naval Service and Air Corps had recommended this. It also called on the department to commission an independent air operations safety audit on behalf of the Air Corps. And it said that the GOC Air Corps should institute a series of operational and equipment-related reviews.

The report said that a formal service-level agreement should be put in place between the Air Corps and the Irish Coast Guard, and between the Air Corps and the airport authorities. Significantly, it said that the GOC Air Corps should review the shortcomings of the

Dauphin as a search and rescue 'platform of operations' in the north-west theatre.

The report drew an immediate call for the GOC's resignation from Tony Baker, the father of Captain Mick Baker. Either he should stand down or he should be dismissed, Mr Baker said. Two government reports had identified the need for medium-range helicopters for search and rescue on the Irish coastline, but these recommendations had been ignored, Mr Baker said. At the time, Mr Baker was unaware of the GOC's memo of 15 June which had advised against initiating the 24-hour base on 1 July 1999.

The GOC welcomed the report as 'very thorough, balanced and detailed' and said that his immediate response would be to 'study and assimilate the content of the recommendations and to implement them with some urgency'. He said that some action had already been taken, in that a flight safety office, headed by a highly qualified flying officer, had been provided.

The GOC gave what some observers would describe as a disastrous interview on RTÉ television that evening. He appeared to demonstrate a complete lack of understanding of the nature of search and rescue, and said he was 'not certain' as to why there was no licensed air-traffic controller in Waterford Airport on the night in question. While such a function was important, he said, it was not always used, given the response time required in an emergency rescue mission. Back-up for the south-east base was a combined responsibility, and there had been many discussions with the Department of Defence and Waterford Airport. 'The back-up must be there, and the back-up wasn't there.'

Recriminations would continue for some time, even as the warnings expressed by those serving and by former Air Corps staff about the Dauphin fleet were to prove justified. Waterford Airport found itself defending its own handling of the relocation, including the air-traffic control issue and airport lighting. However, the Department of Defence and its relationship with the Air Corps became the immediate focus of attention. The *Star* newspaper on 7 October 2000 claimed that a report prepared by Air Corps personnel for the military authorities suggested that crash rescue services at Air Corps installations were inadequate.

The minister called a briefing and issued a statement in which he confirmed that he had been told about the lack of a fire tender at Finner. The fire tender was scheduled for delivery early in the new year and a new tender was also on order for Baldonnel. The minister said he had 'impressed on the GOC Air Corps that safety standards in both equipment and facilities are of paramount importance and that funding is available to improve and modernise safety equipment where necessary'. He said he had been assured there were no other safety questions of which he was not aware. He said he had emphasised to the GOC that any Air Corps operations where the GOC was not satisfied with safety standards in place should 'be suspended immediately'.

The investigation was raised during priority question time in the Dáil on 24 October 2000, with two questions tabled by the Fine Gael deputy, Michael Finucane, and Labour deputy, Jack Wall. A row erupted after the minister's statement on the report, in which he said that all safety recommendations set out in the report which called for action by the Department of Defence and the Defence Forces were being considered as 'a matter of priority with a view to their immediate implementation'. Finucane said that the minister's first words should have been an apology to the families of the four men who died. 'This tragedy should not have happened. They should not have been placed in that situation and the Minister should ask serious questions within the Department of Defence and of [the Department of the] Marine and Natural Resources.'

Pressed further by Finucane, the minister said: 'To suggest there was not the preparatory work of the most painstaking kind undertaken by the Air Corps and officials in the Department of Defence is to deny a fact which Deputy Finucane knows.' When Labour's Jack Wall intervened to ask why there was no air-traffic controller on duty at Waterford Airport on the night in question, the minister said that the airport manager was there, and there was a technician. 'However, it is important to note that it is always the captain, the pilots and the people in the aircraft who make the judgment call on the night. It is not me. . . .'

Reading the transcript of the Dáil proceedings, the bereaved families were deeply upset at the minister's remarks. It was to mark the first shot in the State's subsequent battle to deny liability for the crash.

On 3 November 2000 a married man with two children fell overboard from the Rosslare–Fishguard passenger ferry some 17 miles off Carnsore Point. Mike Davis, a 25-year-old engineer with Kilkenny County Council, was in the water for an hour and 22 minutes before the Dublin-based Sikorsky helicopter arrived. The Air Corps Alouette at Waterford had offered to fly, and there were claims that Davis might have survived if it had been sent. However, conditions were misty, and there was some question as to whether the ship had done a 'man overboard' drill. Later that month night-time helicopter cover from the Air Corps base in Finner had to be suspended because of technical difficulties with lights and a winch on the Dauphin.

At this stage the lack of crash rescue facilities at Finner and several other military bases had become a major issue between the Air Corps and the Department of Defence. It emerged that the absence of these facilities in the north-west had been identified over five years before in a report to the GOC. Pending provision of a fire tender at Finner, the helicopter was moved to Sligo Airport.

Maria O'Flaherty, widow of Captain Dave O'Flaherty, broke her silence that same month. In an interview published in *The Irish Times* on 6 November 2000, she said the four men had been badly let down by their employers. The Department of Defence had a lot to answer for, and the official response to the investigation findings had been 'completely inadequate'. Although the minister had committed himself to purchasing medium-range helicopters as a replacement for the Dauphin fleet, the tendering procedure would mean a delay of at least three years. 'They should be leasing medium-range aircraft in the interim. She said she found the official report to be comprehensive.

One of the most shocking aspects, for her, was the revelation that her husband and crew were not informed until 1 July, the day the 24-hour base was initiated, that no after-hours air-traffic control was being provided. The Air Corps had requested it, but there had been no financial agreement between the airport management and the Department of Defence. Air Corps senior management had told the investigators they were 'unaware' that the issue had not been resolved. The Department of Defence officials dealing directly with the Waterford deployment stated that 'they saw it as a matter of fact that after-hours local air-traffic control services would be provided'.

'This sort of answer is just not good enough,' Ms O'Flaherty noted. 'It is evident that facilities were not adequate at Waterford, and someone in the Department of Defence wasn't doing his or her job.' Ten days after publication of her interview, comrades of the four Air Corps men called on the government to implement all 25 safety recommendations in the investigators' report. The president of the Representative Association of Commission Officers, Lieutenant Colonel Paul Allen, said the average age of the Air Corps' 32 craft fleet was 24 years, and most of these craft were older than the pilots that flew them. The organisation's general secretary, Commandant Brian O'Keeffe, told the *Irish Independent* on 13 November 2000 that those responsible — whether they be within the Air Corps, the Department of the Marine, the Department of Defence, the government or elsewhere — must be held accountable for the systemic failures that contributed to the Tramore crash.

He said the report clearly illustrated the potentially disastrous effects on an organisation and its members of substituting, cherry-picking, cost-cutting, split authority and micro management of what should be an integrated, strategic approach. Air Corps members confirmed that the multi-tasking role expected of the organisation over the years had placed great strain on it. Dauphin crews had long expressed concern about the demands of combining search and rescue, ministerial air transport, security and other duties, but these concerns had been ignored.

The inquest, which had opened in July 1999 without the knowledge of most of the relatives, was resumed on 21 November 2000 in Waterford. Acting coroner for east Waterford, Dr Eoin Maughan, said that having read the official report, he was satisfied that it had no bearing on the matters to be addressed. Dr Joseph O'Connor, who carried out the post-mortems, found that all had died of multiple severe traumatic injuries with subsequent conflagration, consistent with having been in a helicopter crash. The jury returned a verdict in accordance with the medical evidence.

Extending sympathies to the relatives, Dr Maughan said they were brave men who did not think twice about going out in terrible weather conditions to help those lost at sea. Afterwards the Baker family issued a statement in which they accepted the verdict and acknowledged that the function of the inquest was not to determine culpability.

However, they were 'firmly of the view' in the light of the recently published official report into the accident, that the question of culpability and the taking of steps to avoid a repetition, were matters which should be urgently redressed. 'While we accept that a resolution of these matters will not bring Mick back, it is essential that they are fully addressed in the interest of helping our family cope with our loss and in the public interest,' the statement said. 'We now intend to direct all our efforts towards progressing these matters, the resolution of which we believe will provide some basic comfort to us as a family.'

In February 2001, Vincent Byrne, father of Corporal Niall Byrne, also expressed his anger. Little had changed since the crash, he said. Air Corps crews still flew rescue missions in Dauphin and Alouette helicopters, and there had been a number of 'incidents' in the Dauphin craft since Tramore.

Rescue cover at Waterford was also to be given to a civilian company, he said. This was confirmed by the Department of the Marine, which said it had to meet the commitment to restore 24-hour cover in the south-east. Contracting out the service would fill the gap until the Air Corps got its new equipment. The Minister of State for the Marine and Wexford TD, Hugh Byrne, denied that the Air Corps search and rescue service was being 'downgraded' and said it was his wish that the defence wing would return to the area when suitably equipped.

In July 2001 the findings of a safety audit of the Air Corps and its craft were published. The audit had been commissioned by the Minister for Defence in the wake of Tramore, and was carried out by a team from Intercontinental aviation safety consultants in Columbia, Maryland, US. The audit team said it found it remarkable that the Air Corps had been able to achieve such a high level of performance in spite of the numerous and significant obstacles that had the potential to 'negatively impact' on safety.

Among the issues it highlighted was the continuing loss of skilled pilots and experienced aircraft technicians, which was leading to increased workloads for those remaining; and the deteriorating runway conditions at Casement aerodrome in Baldonnel, the Air Corps headquarters. It found one runway which was in very poor repair, noted that concerns over the condition of the runways had been raised as far back as 1993, and warned that a plan to carry out improvements only in 2003 was 'ill advised from a safety perspective, and not acceptable'.

The wait for the promised new equipment would be even longer than Maria O'Flaherty had anticipated. A tendering procedure initiated by the Minister for Defence became highly controversial when politicians lobbied in favour of one bidder, the US manufacturer, Sikorsky, since it was claimed that an offset deal would save up to 1,500 jobs at the north Dublin FLS Aerospace plant. A review group examining the tenders for between three and five medium-lift aircraft was believed to have favoured a rival bidder, Eurocopter.

When the €87 million contract was awarded to Sikorsky in

March 2002, Eurocopter issued a legal challenge. On 5 July 2002, just over three years after the Tramore crash, the minister announced that he was cancelling the deal as part of a €40 million cut in defence spending. Some €12.5 million had already been put aside to buy the helicopters.

The minister described the decision as 'painful', and said the main priority for the Air Corps was the purchase of fixed-wing training aircraft. Meanwhile CHC Helicopters was awarded a contract to provide 24-hour rescue cover at Waterford. There were several start-up delays because of problems in getting crew. Several weeks before the final promised start-up date of 1 January 2003, Waterford Airport ran into financial difficulties. When these appeared to be resolved, a new target date of June was set.

The military court of inquiry finished its deliberations on 30 August 2001. Normally such courts are confidential, but copies of this report were given to the families, with several sections left blank. The court concurred with most of the AAIU findings, but had additional details — on attempts to obtain a weather forecast, for instance. It identified 'serious organisational inadequacies' with the Dauphin helicopter service launched at Waterford on 1 July, and also identified inoperative facilities at Waterford Airport on the night in question.

The relatives were upset at the blank sections; a Defence Forces spokesman told *The Irish Times*[5] that the report was for internal use, but it had been agreed to give copies to the relatives. Sections had been deleted because military witnesses had given evidence relating to Waterford Airport and personnel working there, the spokesman told the newspaper. Some Air Corps and ex-Air Corps colleagues were also disturbed by what they believed to be an attempt by the military court to implicate the pilots.

In a section on the actions of the detachment, the court found that the Air Corps technical support and the winching crew had not contributed to, and were not responsible in any way for, the crash. In the case of the pilots, the court said it was mindful that neither had survived to give their version of events; both were clearly motivated by a strong sense of duty and commitment; and both had been aware that one of the casualties in distress on the small boat was a child. However, it found a number of 'errors of judgment' and failure to follow certain procedures which were 'contributory causes' to the crash, including the failure by the detachment commander to notify Air Corps headquarters when he became aware that no air-traffic control officer would be provided outside of Waterford Airport opening hours; and his failure to obtain an up-to-date weather briefing personally. The pilots 'ought to have known' that the flight had taken off in conditions which were at or below the prescribed minima for landing, and during the mission the pilots had failed to make regular weather and 'operations normal' contacts with base via VHF radio.

Yet the report also found that the pilots' actions must be considered within 'the context of the pressure they were under' and must also be assessed within the context of 'serious organisational deficiencies' and the 'failure to provide them with the best systems of operation that could reasonably have been provided in the circumstances'. It found the absence of a qualified air-traffic controller on the night in question to be a 'very significant contributory cause of the crash'.

The report concluded that no single factor had caused the crash, and that it was the result of a chain of events. It said it was struck by the 'genuinely high regard' in which the four deceased had been held. And, in a most damning paragraph, the court said it was of the opinion that in view of the level of activity at Waterford Airport on 1 July 1999 and the resulting pressures associated with the initiation of the new Dauphin service, 'that same crew should not have been available for call-out on that night'.[6]

The report was signed off in August 2001, even as references to 'pilot error' appeared in several press reports referring back to the Tramore crash. Coincidentally a letter appeared in *The Irish Times* that month from Captain David Courtney, chief pilot with the Irish Coast Guard's Shannon helicopter and a former Air Corps pilot, which effectively put that claim to bed: 'In aviation, there is an area of study devoted to human factors in accidents, known as crew resource management (CRM). This subject examines the complex behavioural and personality traits and patterns that we humans exhibit, both in normal activity and, more importantly, in emergency situations. It is a compelling subject. Its incorporation into worldwide pilot training academic syllabuses, civil and military, has already made a positive impact on flight safety by reducing accident rates. One of the key issues in CRM is the avoidance of misusing the term "pilot error". Accidents happen as a result of a chain of events catastrophically occurring in sequence. The breaking of this chain either stops the accident, or may reduce its seriousness. The human element in that chain, the pilot, is but one link.

'In the case of the Tramore Dauphin crash of 1999 . . . the Air Accident Investigation Unit of the Department of Public Enterprise produced an exhaustive report detailing the chain of events that led to the loss of four lives. It is essential to read the entire report, not just extracts, and observe the chain.

'Search and rescue (SAR) helicopters, especially those which operate at night, perform a task not fully appreciated or understood by fellow aviators, let alone the general public. SAR crews frequently carry out rescue missions at night, at maximum range, in appalling weather, sometimes over mountainous seas, when they are themselves tired. They may have been wakened from their sleep and told that other people's lives depend on the success or failure of their actions. There are other dynamics in a rescue such as language barriers, weather changes, systems failures on board the ship in distress, systems failures on the helicopter itself; the list is endless.

'Where non-rescue pilots avoid putting any links in place that could lead to an accident, SAR helicopter crews operate on occasions with such links in place, especially on night missions. Hence the term, "operating on the edge". Why else do you think they wear layers of thermal suits under dry diving suits, as well as carrying emergency oxygen diving cylinders? In case of ditching, that's why.' (Annual emergency underwater training is compulsory for Air Corps and Irish Coast Guard SAR helicopter crews.)[7]

A memorial to the four aircrew had been erected in Tramore on the first anniversary of the helicopter crash — a stainless steel sculpture by Waterford-based artist John O'Connor. The formal unveiling would take place several months later. The artist explained that the design was based on the technical elements of the helicopter which had crashed near by. It included four rotors, similar in size but slightly different to reflect the individuality of each crew member. Four fibre optic lights would shine from the structure at night, and the rotors would catch the wind — 'reflecting the idea that the memory of these brave men will live on,' he said.

The Air Corps helicopter wing paid its own personal tribute to the four colleagues at their 'home from home' in Donegal. A bench cut from limestone, commanding a clear view of Slieve League, St John's Point and Donegal Bay, was dedicated to the four just outside the search and rescue base at Finner camp on 12 November 2000. They had spent much of their flying life at the base. Finner's association with the Air Corps spanned many decades and the first Alouette had been deployed there in 1973 for border duty, owing to worsening political strife in Northern Ireland.

Members of the Army 28th Infantry Battalion at Finner and over 300 representatives of the south Donegal community, the fishing industry and marine and mountain rescue volunteers heard Lieutenant Colonel Aidan Flanagan, Officer Commanding the helicopter wing, describe the sense of loss felt by colleagues and friends of Rescue 111. Lieutenant Colonel Declan O'Carroll of the 28th Infantry Battalion recalled the devastation at the north-west camp when news of the accident in the south-east first came through. Army chaplain Father Alan Ward noted that the men had been described as heroes after the fatal accident; but they had been heroes before that, he said, in the work they did and in the way they had led their own lives at home.

Air Corps and Army personnel, Donegal County Council, the Killybegs Fishermen's Organisation and individual fishing skippers and managers had contributed to fund-raising for the bench, which was carved by a Creevy stonemason, Joe Roper. Co-ordinators of the project were Commandant Sean Clancy and Commandant Tom O'Connor of the helicopter wing. There were no politicians and not too many public figures on the invitation list.

The wind was a biting north-westerly as a commemorative plaque was unveiled by two widows and two mothers — Maria O'Flaherty, Monica Mooney, Mary Baker and Anna Byrne. The bench had been finished with a four-line extract from W. B. Yeats's poem, 'An Irish Airman foresees his Death', his beloved mountain, Ben Bulben, providing the backdrop just over the county border. The quotation had been selected by Captain O'Flaherty when he was associate editor of a publication marking 30 years of Air Corps helicopter operations in 1993:

> Nor law, nor duty bade me fight,
> Nor public men, nor cheering crowds,
> A lonely impulse of delight
> Drove to this tumult in the clouds. . . .

15

FLYING BLANKETS AND THE BIG MAN FROM ARMAGH

Emily Hackett was an experienced and highly respected climber. On Sunday, 10 October 1999, she was out with friends from the Irish Mountaineering Club on a sea cliff route on Ardmore Head, close to the border between Cork and Waterford.

'It was an easy route and one which I had put up myself,' she says. 'I feel I must have slipped on the grass at the top or else put my foot on wet rock as I was climbing, but I honestly don't remember much about what happened.'

Emily fell 150 feet down on to a ledge and was seriously injured. The alert was raised by fellow climbers. Captain (now Commandant) Shane Bonner, Sergeant Ciaran Murphy and Corporal Aidan Thompson were on duty at Waterford Airport; an Alouette had been deployed there after the fatal Dauphin crash to maintain daytime air-sea rescue cover in the south-east.

The local coast and cliff rescue unit, a doctor and the RNLI lifeboat from Youghal in east Cork were at the scene when the Alouette arrived shortly after midday. The cliff rescue unit had reached her, but she was too badly injured to be moved. The only option was to winch her off in a stretcher by helicopter. She had already been covered with a thermal blanket to keep her warm.

The helicopter decided to try an approach; Bonner flew in towards the ledge, and Aidan Thompson was already out on the winch cable when the unexpected occurred. The thermal blanket was whipped off the injured woman in the aircraft's downwash.

There was a bang, the helicopter started vibrating violently and there was a loud clicking noise. Bonner had seen the blanket fly up, but there was nothing he could do so close to the cliff and with Thompson out on the cable. 'We were sinking, even though I was using all the power I had available,' he says. 'I shouted to Ciaran

Murphy to winch Aidan in.'

At first he thought the blanket had hit the helicopter's tail. 'In fact it had hit the main rotor. If it had been a tail strike, we would have lost directional control and started to spin.' The aircraft was still losing height even though the pilot was applying full power. He moved out left away from the rocks and towards the sea, as Murphy worked hard to winch Thompson back up. Bonner opened the flotation gear switch-guard. He was just about to press the button to deploy the helicopter's emergency floats and call to the crew to prepare for an emergency ditching, when the aircraft levelled off at about 40 feet.

The winchman was still out on the cable and within 30 feet of the aircraft when the pilot slowly climbed away. Once Thompson was back on board, Bonner landed on a nearby cliff top. The aircraft was still vibrating and the loud clicking noise continued throughout. 'We landed and shut down.'

The crew inspected the helicopter, but the blanket had vanished and they couldn't see any damage. 'One of the coast and cliff rescue service was driving a four by four and he headed up to us,' Bonner says. 'We told him we didn't think it would be safe to continue as we couldn't guarantee the aircraft was safe. He suggested calling the Sikorsky from Shannon, and he radioed back to his colleagues who were with the casualty.

'The word that came back on the radio was that they didn't think she was going to make it as she had very serious head injuries. She needed to get to hospital immediately. We had a quick chat among ourselves; I told the winch crew there was no pressure on them if they felt they couldn't continue. But we all knew that Shannon would take an hour to get there, and the lifeboat couldn't get near her as she was 50 feet up. We decided we'd start up, do some hover checks and see how we got on.

'So we did start up, took off and spent five minutes doing the hover checks. There was no clicking noise at this stage and everything seemed to be working normally, so we decided to go for it and head down towards the casualty. However, we called Shannon as we were airborne, and the S-61 was scrambled to follow us in case we got into trouble. Thompson went down on the cable, hooked the stretcher on and Murphy took them both back up.

'We flew directly to Cork Regional Hospital, but it certainly wasn't a very pleasant experience. We were very wary and listening out for any unusual noise, and the winch crew were working hard all the time to get life signs from the casualty. Twice they asked me to land in a field as they thought they had lost her, and as we descended, each time they managed to get her back. We landed at the hospital, and the Shannon helicopter was five minutes behind us as we handed the casualty over to the hospital personnel.'

And where was the missing blanket? 'The engineer had three theories,' Bonner says. 'Either it went in and blocked an air intake, which would explain the loss in air power; or it was sucked into the engine; or part or all of it got wrapped around the main rotors. I think myself it was probably the third possibility, as that would explain the loud click. I imagine it just got shredded into pieces then. We were all trained on emergency ditching procedures at the Royal Navy base at Yeovilton, but I suppose it's always better not to have to put that into actual practice!'

The three crew had made a very difficult decision to continue the mission — and within three months of the Dauphin crash at Tramore — but they knew that Emily Hackett's life depended on them. The details were reported to the Air Accident Investigation Unit as a matter of course. In its findings, it said that the incident re-emphasised the need for joint training between helicopter operators and other agencies involved in search and rescue. 'Ground-based personnel in these circumstances must be constantly aware of the power and consequent dangers of helicopter downwash.' It said that lightweight items must be 'removed from the helicopter's working area, or if necessary must be strongly secured. In this instance there were some ten personnel of various agencies around the casualty. This precluded the helicopter crew from seeing the casualty before the aircraft moved in to lower the winchman.'

It also recommended that the Irish Marine Search and Rescue Committee, which involved representation from all the rescue agencies, carry out a review of the details of this training by marine rescue teams, the gardaí and mountain rescue teams. And it said the committee should look also at the 'command and control' procedures at rescue sites.

Emily Hackett was treated at Cork Hospital and made a full recovery after rehabilitation. 'I am for ever grateful to the Air Corps for my rescue as I know it was not easy for the crew, when it was not long after the Tramore accident, and especially when the space blanket got caught in the propellers,' Emily Hackett says. 'If anything had happened to the crew, that would have been a very difficult thing for me to live with.'

Late that year Ms Hackett received a phone call from the Air Corps to say they had her watch. 'I was doing a course at the National Rehabilitation Centre in Dun Laoghaire at that time. While there, I saw a helicopter land. I went out to look at it. They told me they were collecting things to bring to the south-east base, I think. I could not believe how little room there was in the helicopter. I must have mentioned my watch, and they remembered, because I received it back the next day at the centre. I couldn't believe it as it was old, but working.'

Shortly before Christmas of that same year, Captain Bonner was involved in another mission which also proved to be very challenging, but didn't have such a fortunate outcome. He was on duty on the Dauphin helicopter, this time in the north-west. On 8 December 1999 the Finner crew took a telephone call from the gardaí in Donegal. Paul Chapman and his son Sean had gone on a short drive in Mountcharles, Co. Donegal, and were travelling along the coast road when a freak wave caught them and washed the car and part of the roadway into the sea. The body of the man was found close to the car but the young boy was nowhere to be seen. The family had only recently moved to the area from England and Mr Chapman had intended to build a house there.

Bonner remembers that the wind at Finner was gusting to above 60 knots, which was just on the limit for starting the helicopter rotors. 'We had to discuss whether we would go or not, it was that bad. We took off and routed to the area, and the conditions in the search area were worse. The wind was 70 knots and gusting higher to around 90 knots. As it was north-westerly and we were on the lee side of the Donegal Mountains, there was also severe turbulence.'

With Bonner were Captain Brendan Jackman, Sergeant Ciaran Murphy and Corporal Benny Meehan. 'Searching conditions were hampered by very heavy showers and the fact that the sea was extremely rough. In fact it was pure white — there was no sea as such,' Bonner says. The helicopter stayed out an hour; there was no sign of the boy. As the storm deteriorated further, the crew witnessed two large lightning bolts, both quite close to the aircraft. The turbulence was making them nauseous. They were forced to return to base.

They knew in their hearts that the young boy would probably not have survived long in those conditions, but they were airborne again at first light. Flying over a beach at Murvagh, they saw the body. The aircraft landed, they took the boy in a stretcher into the aircraft and flew him to Sligo Hospital. He was pronounced dead on arrival.

It was a particularly sad journey for the crew, most of whom had family themselves. Bonner, from Monaghan, has worked on search and rescue since 1995. 'You don't do this job for ever. But I told my wife I would stop only when I no longer got that adrenalin rush on hearing the alarm going off.'

ONE MAN WHO HAS had more than his share of adrenalin rushes is the CHC Helicopters winchman Noel Donnelly, formerly of the Air Corps and now working at the Irish Coast Guard base in Shannon. On 31 January 2000 several Spanish 'flag of convenience' fishing vessels were working off the Irish west coast when a fire was reported on board one of them — the *Milford Eagle*. The vessel was 150 miles west of Shannon when it issued a Mayday at 1.46 a.m. At this stage the 17 Spanish and Portuguese crew on board had been unable to control the blaze and had taken to liferafts, which they wisely lashed together.

The Irish Coast Guard scrambled an RAF Nimrod, Rescue 51, from Kinloss, Scotland, and both the Shannon and Dublin-based Sikorsky helicopters. Pilot of the Shannon helicopter, Rescue 115, Captain Derek Nequest, was airborne at 2.54 a.m. He remembers it was a misty night with strong winds, big seas and poor visibility. The RAF Nimrod was the first to locate the burning vessel, having spotted a flare. It reported the position to the helicopters, which were battling against 60 knot headwinds. The Nimrod then sought assistance from other vessels in the area, and the Irish-registered Spanish vessel, the *Alimar*, reported it was within a 15 mile range. Ironically, both the *Alimar* and the *Milford Eagle* had been apprehended just over a week before by the Naval Service for alleged fishing offences in Irish waters.

With Nequest were co-pilot Captain Robert Goodbody, winch operator John Manning and winchman Noel Donnelly. Manning used the aircraft's infra-red heat-seeking equipment to detect the liferafts, and after a quick reconnaissance it was decided to let Donnelly down the cable.

'The winch operator can normally fly the aircraft using a joystick in the back, and if you are hovering over a liferaft that is sometimes the best way to go about it. He can see, and position the aircraft accordingly. The authority that he has, using that control, is not very great, and on that particular night the sea conditions were such that it wasn't going to work,' Nequest says.

'So we were faced then with a situation where the pilot, that is me, trying to hover the aircraft over something he can't see. On a flat calm sea it is feasible, but on a big heavy sea where you've got this goldfish effect, it is particularly difficult, and of course winching is very different at night-time. Lack of vision, lack of horizon . . . makes it far more of a challenge and requires a different technique.'

Disappearing into the black night, Donnelly thought he was descending to assist a small crew. 'Initially on take-off we were told there were five on board and they were preparing to abandon the vessel. And then about a half-hour later we received confirmation that there were five people in liferafts,' says Donnelly. The big Armagh man had to swim to the liferafts, with the cable still attached to him, in 20 to 30 foot seas. To his surprise he found 17 in total. 'We knew we couldn't pick them all up in the time we had. So I sent up four, two at a time, and said I'd stay with the rest. A few of the younger crew were panicking a bit and some had minor burns on their feet, so the skipper had prioritised who should be taken off first.

'We had passed another fishing vessel on the way out, and it was within 20 miles. It took a while to get it to respond and I think the RAF Nimrod had to wake it up! Four or five hours later it came and picked us up out of the liferafts, and then the Dublin helicopter winched us off from there.'

Donnelly makes light of the time he spent with the terrified Spanish and Portuguese crew during those hours till dawn. The RAF Nimrod remained at the scene and provided a vital communication link. 'None of the crew had lifejackets. They were all in normal clothes as they had obviously left the burning boat in a hurry,' Donnelly says.

'One or two of the 13 still with me then had workable English. I had to try and explain that if any of the rafts capsized they should hang on to it, and we would right it again and get them all back on board. At this stage the sea was building to 30 to 40 foot waves, but luckily there were no capsizes. I just kept hopping from one liferaft to another all the time to make sure everyone was OK.'

'Noel did a great job,' Nequest says. 'Of course we knew that although we were abandoning him, he was in no real danger because he had his hand-held radio — through which we could make reassuring noises! There was another vessel close by at this stage, and the RAF Nimrod above. By the time we left we could see the *Alimar*. It picked them up off the liferafts. Meanwhile the Dublin aircraft had been called out. This wasn't unusual, and had happened before. If it looks like we are going to have a long job on the west coast, the Irish Coast Guard will call Dublin out. And that helicopter actually picked up the remaining survivors from the *Alimar*.'

Donnelly says that the skipper of the *Alimar*, Captain Alfredo Garcia, applied considerable skill in coming alongside the liferafts in such big seas, and providing shelter while the crew were taken on board. Within 15 minutes Donnelly and his survivors were back up on deck as the Dublin Sikorsky — flown by Captain Gordon Baird, co-pilot Hayden Lewis, winch operator Steve Dodd and winchman Paul Ormsby — prepared to winch them off.

Donnelly remembers that one of the men had to be taken up in a stretcher; at this stage he was in hypothermic shock. 'We were all flown in to Shannon, landing shortly before 8 a.m., and the survivors were taken by ambulance to hospital in Ennis and Limerick for medical checks.'

Later that day an Air Corps Casa fisheries patrol plane and the Naval Service patrol ship LE *Aisling* were dispatched to locate the position of the *Milford Eagle* and ensure it posed no threat to the marine environment. The vessel had been drifting in a north-easterly direction when last sighted, and its insurers commissioned salvage assistance to tow it into Killybegs, Co. Donegal. The patrol ship LE *Deirdre* later took the abandoned vessel in tow and transferred it to its Spanish owners.

The rescue crews, staff at the Marine Rescue Co-ordination Centre in Dublin and the Valentia and Malin Coast Guard stations were commended by the Minister for the Marine, Frank Fahey. The minister said that Noel Donnelly had displayed 'tremendous courage' and he also wished to thank the crew of the RAF Nimrod and the Irish-registered Spanish vessel, the *Alimar*. Also commenting on the rescue, the Fine Gael MEP for Munster, John Cushnahan, said the EU must shoulder more of the cost of providing search and rescue off the Irish coastline, as emergency services were stretched to their limit. He also repeated his call for the establishment of an EU Coastguard which could deal with such situations and share the financial burden.

Rescue 115 was subsequently given a letter of appreciation for meritorious service at an awards ceremony attended by the Minister for the Marine in Dublin Castle in November 2002. Similar letters of appreciation were presented to the captain and crew of the RAF Nimrod, Rescue 51 — accepted on their behalf by Lieutenant Neil

Eccleshall — and the skipper of the *Alimar*, Captain Alfredo Garcia. His award was accepted by the vessel's Irish owner, Brendan Rodgers. And for his 'extreme courage, skill and determination, risking his life to save others', the Michael Heffernan Silver Medal for Marine Gallantry was awarded to Noel Donnelly.

16

SKERD ROCKS SURVIVOR

Ricardo Arias Garcia (24) was awake in his bunk in the storm. It was a severe gale force nine blowing to ten, and his fishing vessel was heading for shelter off the Irish coast. The 32 metre *Arosa*, Spanish-owned and registered in Milford Haven in Wales, had been trawling for about four and a half days and had hauled the gear shortly before 7 p.m. on 2 October 2000 when the weather worsened.

The *patrón de pesca* or skipper, Ramon Pardo Juncal, had decided to head for Galway Bay at a full speed of ten knots. Juncal was just 31, a young but relatively experienced fisherman, having spent more than a decade working on the Gran Sol bank some 100 miles south of Fastnet, and off the Falkland Islands and South Africa. He had been with the *Arosa* for over a year, and this was his third trip as skipper. Working with him were twelve crew — nine Spanish, two from Sao Tome Island off Africa and one from Ghana.

At 9.42 p.m. the vessel made a compulsory call to Valentia coast radio in Kerry, stating it was leaving one fishing area and moving into another. It also reported it had four and a half tonnes of fish on board. By this time the crew were sitting down to dinner — a bumpy meal as the vessel heaved and chucked in the heavy seas. Afterwards some showered and went to bed, like Ricardo, and others stayed in the messroom. The *patrón de costa* or first mate, Euginio Diaz Carracelas, was up in the wheelhouse on his own. It had been a long day and he wanted everyone to have a good rest.

Ricardo was unable to sleep. Shortly before 4 a.m. he heard a loud bang — another large wave hitting the side of the vessel, perhaps. It was followed very quickly by a sickening, grinding groan as the *Arosa* struck rocks, its engines still running. The alarm sounded, and he heard the voice of the *patrón de costa* roaring at them to come up on deck because the vessel was taking water. Within minutes he and his fellow crewmen were scrambling up, no time to grab clothing, no time to grab anything other than each other as the skipper dashed into the wheelhouse and sent a distress message by radio.

The vessel had run up on the fearsome Skerd rocks about ten miles north of the northern entrance to Galway Bay and nine miles west of the village of Carna. Huge seas were crashing over the jagged teeth as the terrified crew grabbed lifejackets and tried to tie them on. There were no survival suits on board. With the vessel listing to starboard, wedged between two stags, one of the deckhands shouted that the liferafts should be launched.

Quick-witted, the three Africans inflated them, but it was almost impossible to board them from the port side. The seas were too fierce. Perhaps it was better to cling to the vessel. Those who tried were washed away; and those who held on were washed away also, one by one. Somehow, Ricardo managed to hold on to the hull. He had no lifejacket on as he feared it might choke him. 'There were seven of us,' he said afterwards. 'With every wave that came, we had to hold on very tight but some just couldn't.' His last sight of the three Africans was as they headed for the ship's bow. 'Then I was the only one holding on and a huge wave came and swept me away too.'

He felt he would not survive. 'In between the waves, I tried to look up, calm down and organise myself. I saw another big wave coming. I closed my eyes and took a deep breath. When that wave had passed, I felt rocks beneath me. I dragged myself up along the rocks. I looked up and I saw the light of the helicopter. . . .'

THE IRISH COAST GUARD helicopter at Shannon had barely returned from another rescue mission when it got the call. Two days before, on 1 October 2000, a distress signal was picked up from a satellite emergency position-indicating radio beacon (EPIRB) attached to a French-registered fishing vessel, the *An-Orient*. It had been relayed from an aircraft through the French satellite system to the Coast Guard. The position of the 38 metre trawler was 87 miles west of Loop Head, Co. Clare. No distress message had been transmitted.

Westerly winds were gusting to 60 knots with a very rough sea and swell building as the Sikorsky battled its way out from Shannon. 'We went out to an EPIRB location, but there was nothing to be seen,' Captain David Courtney, the pilot, says. 'We were facing into this 60 knot wind and looking around, and then we saw streaks on the sea surface that looked like oil. We approached from about two miles downwind of the location to start searching and saw three of them after the second sweep. I spotted them and then I remember Noel Donnelly on the infra-red saying "I have them. There they are." They were in the water holding on to a life-ring.'

The three, including the French skipper, had been in the water for about four hours and were suffering from hypothermia and shock. 'It was fairly rough, but it was daylight so it was a textbook operation,' Peter Leonard remembers. 'There was quite a heavy swell, but it was manageable. The boat had sunk, and we were listening out for the EPIRB. As we picked it up, we flew over wreckage and some debris. We circled around, and went upwind as the debris was drifting with the wind and tide, and then we spotted three people in the water. The three were hanging on to two lifebelts between them. The youngest of the three was weak and not long from going under, so he was the first one I picked up. The other two were very cold, having been in the water for two to three hours. They were only wearing their clothes, with no survival suits or lifejackets. Talking to them on the way back — because one of them had a little bit of English — they said there had been four of them grouped together, but while we were on the way out there he went under and they lost him. He was a brother of one of the survivors.'

'Peter had to do all the work himself there, because two of them were in a fairly bad state,' Noel Donnelly, winch operator, points out. 'The third man was fairly strong and he was holding the two together. At one stage Pete strained a few ribs because of the effort he had to make.' Courtney concurs: 'Peter Leonard did a tremendous job there. There can be a survivor's tendency to relax too much and die just as they are being rescued — something that was very well described by international yachtsman Tony Bullimore in his account of his five days in an upturned yacht in the Southern Ocean in January 1997.[1] There is so much determination to live that when at last they see a rescue helicopter they can sometimes slip away. Peter had to pull them out in that state.'

They were wrapped in blankets, the heating was on in the aircraft and the crew had a flask of hot water and gave them hot drinks. Courtney and co-pilot Captain Mike Shaw continued to search the area for more survivors, but they knew they had to get the three fishermen to hospital. They headed east, landing at University College Hospital, Galway. Afterwards the vessel's skipper, Xavier Leauté, said he and two crewmen were on deck when a large wave had hit the bow and the vessel listed and began to sink. Built in 1975, the L'Orient-registered trawler had had an extensive refit the year before. It was one of four part-owned by the French supermarket chain, Intermarché.

The skipper said there had been no time to launch the liferafts or to alert the remainder of the crew below, and they had jumped into the water clutching lifebuoys. The ship's bosun, Fernando Neves from Aveiro in Portugal, described seeing the second engineer in the water and swimming towards him. 'He died in my arms.'

The Sikorsky continued the search that day along with an Air Corps Casa maritime patrol aircraft and several fishing vessels. There was no sign of eight fishermen still missing; it emerged that of the eight, six were French, one Portuguese and one was Irish. Tomás Kelly of Fenit, Co. Kerry, had joined the French vessel only the week before, after it put into Fenit, and he was assisting it with new gear. Known as 'Irish Kelly' to his French skipper, he was an experienced fisherman who had part-owned his own vessel until some months previously.

Married with four children, Tomás had been involved with his own community. He had served as a voluntary lifeboat crew member, and Father Gearóid Ó Donnchadha, secretary of the Fenit lifeboat, described how he had participated in the delivery of a Trench class offshore lifeboat to the Kerry station from RNLI headquarters in 1994.

A French fixed-wing maritime surveillance aircraft and several French fishing vessels joined the search area extending to 80 miles west of Loop Head. Apart from some debris on the water, the *An-Orient* sank without trace. An award in appreciation of meritorious service was later presented to the crew of the Irish Coast Guard helicopter for their skill and commitment during the rescue.

The Shannon aircrew were back at work the next day, hoping for a quieter duty. It was not to be. Valentia coast radio station in Kerry picked up a distress message from the Spanish fishing vessel, *Arosa*, in the early hours of 3 October. Its position was latitude 53 degrees 15.21 minutes north, longitude nine degrees 59.42 minutes west. The vessel was taking water and needed a helicopter. The duty crew for Rescue 115, the Irish Coast Guard medium-range Sikorsky, was tasked, as was the Aran Island lifeboat.

A Mayday relay was picked up by the Naval Service patrol ship LE *Eithne*, two Aran Island ferries and three fishing vessels, the *Iuda Naofa*, the *Capall Bán* and the *Endurance*. By this time the weather at the scene was south-westerly winds of 50 knots and visibility of 500 metres. The Sikorsky was called out from Shannon and Captain David Courtney, on duty again, remembers that the most difficult part was navigating there safely.

'We had a lat/long which we had put into the computer, but we knew there were mountains to the north of us, and the Skerd rocks were higher than us at 80 feet. We navigated using the Global Positioning System (GPS) and our radar, and transitioned from cruising over Galway Bay at 2,000 feet to crawling forward at 80 feet and 20 miles per hour. We could see nothing. It was pitch dark and we were in fog. Then we made out the shape of the Skerd rocks on our radar, and we began to see them come into view. Our searchlights showed the rocks were level with us, so we climbed up over the top and saw the liferafts floating, lit by torches which are activated by the sea. Then we saw the vessel immediately afterwards, and so we thought we were going to get them all. But all the liferafts were empty.'

'The weather was dank and the pilots did very well to get out there,' according to winchman Eamonn Burns. 'Then, moving from the planned navigational route into the area was even harder as it took an awful lot of caution. We knew there were small islands, visibility was down to 200 or 300 yards and it was still night-time. Any lights we shone reflected back into the cabin; any water we could see was kind of white with the fog and lights around us. It was like flying in a bubble. The only thing that we knew was presented on a radar. So it was all eyes out, and there was good command from the captain on how we were going to get into the area.

'We started off downwind of a datum area as marked out by the actual EPIRB,' Burns says. 'We took it very gingerly as we knew there were high mountains on the right-hand side. We started to come across and our lights hit the rocks, and then we saw the vessel. It was literally being pounded, and as we caught sight of it we saw the young fellow, Garcia, standing just on the waterline in his underpants and T-shirt. He was just above the bow of the boat, which was semi-submerged, and holding on for dear life to rock.

'Imagine being put into a pitch dark room with no references whatsoever, and you knew that if you moved left or right you could be in trouble. That's what it was like for him. As far as he was concerned he was up against a cliff which could have been several hundred feet high in the pitch dark. It was imperative that we get him out of there. One more wave or a wrong movement and he was gone.

'It was freezing cold, with lashing rain and huge waves. I was lowered down, stumbled over and grabbed him. We picked him off and got him into the aircraft, and did a very quick medical check on him. He was breathing, he was anxious, he was nervous, he was in shock, but he was still alive.

'We did a scan then because we knew there were more crew there. We searched the rocks and saw a liferaft on the far side that was empty. There were life-rings around it which were also all empty. The weather conditions meant that we couldn't move backwards, so we punched into the computer to get us back to the start position. It was pretty hairy as land was appearing on our radar. We checked our maps to make sure the high ground was well away, and we then flew down low to do a visual scan.

'We came across one individual and then we saw a couple of lifebelt lights in the water. The man we had seen had his face up and I think he waved at us. He was trying to maintain his position on his back because of the swell. I went down to him and it took me several minutes to swim towards him as I was still attached to the hoist cable.

'The waves were lifting me up. John Manning was trying to control the cable so that it didn't tangle around me. Every time I hit a wave, he'd disappear. When I got to him, he was still face up, but he was getting washed with water at this stage. We got him back into the aircraft and had to resuscitate him. He had a heartbeat but he was unconscious. We had to do advance treatment on him and try and bring him back, and all the time I was receiving cardiac rhythms. He was responding to pain, and we kept working on him — myself and John.

'Even as we were doing this visibility was very bad, but we thought we could make out more guys with their faces down in the water. It was at this stage that we had to make a decision — this guy still had life signs, so should we try to get him back into Galway which is 20 minutes away, or keep looking? We made a crew decision, based on my opinion, and headed back to Galway and handed him over to A & E. He was the skipper of the vessel. Unfortunately he didn't make it. It's one that I have nightmares about.'

Barely clothed, with a blanket round his waist, his legs gashed in several places, a shivering Ricardo was carried into University College Hospital, Galway, by Eamonn Burns and one of the hospital staff. Some hours later he agreed to speak to journalists from his hospital bed through an interpreter. At this point he did not know he was the only survivor. As far as he was aware, his skipper was still alive, though gravely ill, and the search was continuing for eleven of his colleagues whose bodies were still at sea.

The rescue services recovered four bodies later that day in very difficult weather conditions. Seven of the crew were still missing when the search was suspended as light fell on the evening of 3 October. That night, Ricardo Garcia was transferred to Merlin Park Hospital in Galway and learned that his skipper hadn't made it. The following day he would have to undergo a second ordeal: he would have to identify the remains of four of his colleagues.

The *Arosa* had put to sea from Marin in north-west Spain on 8 September 2000. Ricardo had joined the vessel the previous January. He had worked as a deep-sea fisherman since he had left school eight years before and had served mainly on factory ships in the north and south Atlantic. On this vessel trips lasted an average of twenty days — standard for the Iberian crews who work long and hard for their keep to feed an insatiable Spanish market for fresh fish.

On 22 September the vessel berthed at Ardrossan in Scotland to land the fish caught so far, and to take on stores. But it also had a hole in its hull. Divers contracted to carry out an inspection found a small gash directly below the forward gallows and about one and a half metres below the waterline. The plate around the hole was soft and a local firm was hired to carry out an internal repair.

During the stop-over a British marine surveyor called to the *Arosa* and asked to see the papers for the crew. He detained the vessel because neither the *patrón de pesca* nor the *patrón de costa* had British certificates of equivalent competency to allow them to sail under UK registration rules. The vessel owners found two British skippers who had full skipper's certificates and they joined on 26 September. The detention order was lifted and the *Arosa* was allowed to leave — which it did later that afternoon.

The following day, 27 September, the vessel was *en route* to fishing grounds off the west Irish coast when it called into Killybegs, Co. Donegal. Here the two British skippers headed for home, being no longer under the spotlight of the British authorities. The Spanish skipper and mate took over, but at this point the crew noted that the repair to the hull was not holding up. It was leaking slightly, and so the mechanic put some metal plates together and inserted them between the frames on either side of the leak to make a 'cement box'.

The vessel put to sea again and began fishing off Donegal on 28 September, moving south three days later. Movements could be tracked, as it had to make compulsory reports to the coast radio stations. The skipper was only a quarter of the way through the fishing trip when he decided to make for shelter against a worsening weather situation. Most larger fishing vessels heave to and ride out a storm in the early stages of a trip; it is invariably safer to be at sea in a blow than trying to approach an inhospitable coast. However, the *patrón de pesca* decided to head for the coast.

It may have been that he was worried about heavy weather damage which the vessel had sustained earlier in the year, and he may also have had the repair to the hull in mind. If one of the double plates lifted in the swell, the vessel could be in trouble. While storing fish on the evening of 2 October two of the deckhands, including Ricardo, spotted that the repair was still leaking slightly. It was only the skipper's third command on the *Arosa*; he may not have wanted to take any risks. Either way, he decided to head for Rossaveal in County Galway. The vessel had never been there before, and he did not give any notice of his arrival.

Rossaveal, which was built to serve the Galway fishing fleet, is not a harbour to approach without good charts and some local knowledge. Its entrance is by a marked channel which is also used by ferries serving the Aran Islands. Groundings by vessels unfamiliar with the port are common, and for this reason the harbour master passed a by-law restricting entry to fishing vessels of over 30 metres fitted with a bow thruster.

However, the Skerd rocks nine miles west of the north Connemara coast should be visible to any navigator and are large enough to be shown on any radar screen in spite of heavy clutter. Even if not identifiable on radar, a vessel making a course east-south-east should spot the large waves breaking over Skerdmore, the largest of the group of rocks. Yet in the five cables of distance between Skerdmore and Doonguddle rock 'no action was taken to avoid an imminent grounding' by the *Arosa*, according to the subsequent investigation by the British Marine Accident Investigation Branch (MAIB).

The investigators ruled out technical or mechanical failure. The *Arosa* had been doing ten knots, suggesting it had not been unduly hampered by winds or heavy seas, and the survivor would have noticed a difference in engine noise if one had cut out. The vessel had no history of steering gear failure. And even though the two British skippers had disembarked in Killybegs, the investigators found no evidence to link the level of qualification held by the *patrón de pesca* and the *patrón de costa* to the tragedy.

They suggested that the *patrón de costa* had fallen asleep on watch in the wheelhouse, having altered course just over an hour before. He had chosen not to take a direct route to the northern entrance of Galway Bay. By steering a course a little to the north, he may have felt that the south-south-west gale would hit further along the starboard quarter, making the movement a little easier on the crew, who were trying to rest.

The *patrón de costa* would have been approaching the end of his seven-hour navigational watch at the time of the accident, when the skipper was due to relieve him. Normally he would have had a deckhand with him, but the investigators concluded that he had decided to serve alone and to allow the crew to rest. Tragically there was no watch alarm amid the electronic equipment on board to ensure that he did not nod off.

If he did fall asleep, it would have been through pure exhaustion, because there was no bridge chair in the wheelhouse — only a dining table stool. He would have been standing up for the seven hours on a vessel which was rolling considerably in the beam-on seas. It has been known for fishermen and other watchkeepers to fall asleep standing up, the investigators noted.

In the following days the Aran Island lifeboat, several fishing vessels, the Irish Coast Guard helicopter from Dublin, the Air Corps Dauphin at Finner camp, Co. Donegal, and a French fixed-wing aircraft took part in the search for the missing crew of the *Arosa*. The Naval Service patrol ship LE *Eithne* provided on scene command, and Naval Service divers were called in. However, Commander Eugene Ryan of the *Eithne* found that weather conditions were too rough to approach the fishing vessel which was facing south-east and 'bow on' to the Skerd rocks. 'We can see its foremast, the top of the bridge and funnel,' Ryan told *The Irish Times* on 3 October. 'The seas are white out here, and the trawler is already breaking up. It won't last the night.'

By the next day, six bodies had been found and six were unaccounted for. Relatives flew over from Spain to identify the remains at University College Hospital, Galway; by this stage the ship's agents had advised Ricardo not to give any more press interviews as the official investigation was already under way. At the inquest in Galway courthouse on 6 October, Ricardo confirmed that there had been no look-out on duty, and the first mate had been on his own in the wheelhouse.

A brother of the *patrón de pesca*, Ramon Pardo Juncal, said he disagreed with the deposition taken from the sole survivor since it had criminal implications, and he also objected to the format of the hearing. The coroner, Dr Ciaran McLoughlin, explained that it was not the role of the inquest to apportion blame, and if the relatives wished to make a complaint there were other fora for this. Later that day the relatives were taken to Rossaveal where they met the harbour master, Captain John Donnelly, and those involved in the continuing search. In Carna they attended a poignant Requiem Mass in a local chapel in memory of those who had died.

The church was packed; representatives of the local fishing community attended as a gesture of solidarity. The parish priest, Fr Padraic Audley, said afterwards that the relatives were devastated, and it was the saddest service he had ever conducted. He sincerely hoped the missing bodies would be found to give some peace to the relatives. A week later a body was washed ashore in Carna and was picked up by the Shannon helicopter; an eighth body was found near Gurteen, Roundstone, a fortnight after that.

For days and weeks afterwards members of the Irish Coast Guard units at Doolin and Kilkee in Clare, north and south Aran, and Cleggan and Costello Bay in Connemara searched the shorelines. Debris, including a blanket and fish boxes, gradually came ashore. On 15 May of the following year Naval Service divers recovered human remains from the wreck of the vessel on the seabed off the Skerd rocks. DNA procedures had to be used for identification.

Interpol took saliva samples from the parents of Luis Miguel Vidal Rivadulla of Santiago de Compostella in Galicia. The tests at the Garda Forensic Science unit in Dublin indicated that the remains discovered were 92 per cent likely to be those of their son. On 7 December 2001 an inquest was held in Galway in which the cause of death was given as drowning. Inspector Tony O'Donnell of the Gárda Siochána in Galway said the verdict would be communicated to the family through the Spanish Embassy in Dublin.

In February of the following year King Juan Carlos and the Spanish Government presented an award to the Irish Coast Guard for its work in the aftermath of the *Arosa* sinking. The *Cruz de Oficial de la Orden del Merito Civil* was accepted by Coast Guard officer, Eugene Clonan, on the service's behalf. During Irish marine rescue awards in late 2002 a number of awards for the *Arosa* mission were given to the Irish Coast Guard helicopter crew, led by Captain David Courtney; to Commander Eugene Ryan and the crew of the Naval

Service patrol ship, LE *Eithne*; to the Naval Service diving team led by Lieutenant Dara Kirwan; to Patrick Mullins, coxswain, and his crew on the Aran Islands lifeboat; to Captain John Donnelly, area officer, and team members of Costello Bay Coast Guard unit; and to Gerard Joyce, area officer, and team members of the Cleggan Coast Guard unit.

Tragically no trace was found of the three Africans on board who had been so quick to launch the liferafts. Two — Sebastian Vaz de Almeida and Albertino Herculano da Costa Cravid — had come from the impoverished island of Sao Tome, and the third, Orlando Soares, was from Ghana. It emerged that the Gárda Siochána had been unable to trace the next of kin of the fishermen from Sao Tome.

Máirtín Ó Cathain, a Connemara-based journalist, wrote in the *Connacht Tribune* on 6 October 2000 about the reputation held by *na Sceirde* — the Skerd rocks. Three miles of massive rocks could be navigated safely only in reasonable weather, but the area was the subject of the song 'Amhran Bréagach' which was, as its title suggested, a lyric composed of lies. '*Dún Godail ag sodar aniar le cóir*' in the song was a reference to *Dún Godail* (Doonguddle), the great rock on which the *Arosa* foundered. 'No wind would ever move it. It was the fortress which sank the Spanish fishermen', Ó Cathain wrote, on a night when an angry Atlantic had 'the kind of fury that once greeted the Armada wrecked off the west'. It was one of the worst sea disasters on the Connemara coast in several decades. Within a few months, there was a proposal to base a wind farm on the fearsome rocks, and a wind measurement mast was erected to test suitability.

Two years later, almost to the day, the Skerd rocks almost claimed more victims. A French-registered Spanish vessel, the *Lioran*, was leaving Rossaveal, Co. Galway, in favourable weather conditions when it ran up on the rocks. Radio contact was lost for a time, but the alert was raised and the Irish Coast Guard sent out a rescue helicopter and lifeboat assistance. The vessel with ten crew on board was successfully hauled off and towed back into Rossaveal for repairs. The 38 metre vessel sustained a hole on its starboard side.

The Commissioners of Irish Lights, the body responsible for navigational lights, beacons, buoys and markers around the Irish coastline, said it would not be feasible to erect a warning mark. Sea conditions in the area were such that the rocks would not sustain a light.

Almost two months after the *An-Orient* and *Arosa* sinkings off the west coast, the rescue services were busy again. On the night of 22 November 2000 the *St Gervase*, a wooden-hulled trawler, was preparing to leave Castletownbere, Co. Cork, on a fishing trip. The crew of four sailed shortly after midnight. At around 3 a.m. on 23 November an aircraft picked up a distress signal and alerted the Irish Coast Guard. There were unsuccessful efforts to contact the vessel by radio. The Shannon Sikorsky was tasked, as well as the Baltimore and Castletownbere lifeboats, the Naval patrol ship LE *Roisin* and the Goleen Coast and Cliff Rescue team.

The search concentrated on treacherous rocks near Mizen Head, Ireland's most southerly point. The *St Gervase* had been due to fish for haddock, and the weather was reasonable apart from a heavy swell. The vessel had given no distress signal other than the alert from the emergency position-indicating radio beacon (EPIRB). Unless there had been accidental contact by the beacon with the water, it was a certain sign that the ship was sinking, or had sunk. Initially, according to Valentia coast radio, the latitude and longitude emitted by the beacon was clouded.

The Baltimore lifeboat crew found debris in the water, including fish boxes, and noticed a heavy smell of diesel fuel. The area was close to where the *Celtic Mist*, a yacht owned by the former Taoiseach, Charles Haughey, had been wrecked on rocks in August 1986. Mr Haughey and his crew owed their lives to the quick-witted actions of the Mizen Head lighthouse crew at a time when the lighthouses were still manned.

There was speculation that the *St Gervase* had struck rocks in the swell while on autopilot; either the steering or engines had failed. The skipper, 30-year-old Gary Kane from Donegal town, married to Alexandra O'Driscoll, was an experienced fisherman. The couple had a two and a half-year-old son. Also on board were Jacques Biger (36) from Brittany who had married Gillian Murphy of Castletownbere the previous May, Timothy Angland (30) from Mayfield, Cork, who had been on the vessel for two weeks but had fished for seven years, and Kieran Harrington from Castletownbere, just 18 years of age and a member of the crew for about two years.

In spite of extensive efforts by the rescue crew, no survivors were found. As the hours passed it became a body search. The position of the trawler was narrowed down when a liferaft surfaced, and Naval Service divers located it on its starboard side in 30 metres of water. The body of Timothy Angland was taken from the vessel's wheelhouse, and Kieran Harrington's body was found in Dunmanus Bay. The bodies of skipper Kane and Jacques Biger had still not been found when the official investigation was published in September 2003. It was unable to determine what happened and said: 'Whatever took place on the bridge of the St Gervase is shrouded in obscurity.'[2] In a curious twist of fate, the St Gervase's former owner, Sean Cotter, had lost his life the previous May when he capsized close to port in Bantry Bay.

On 6 November 2000, Irish and British rescue services assisted two Spanish fishing vessels which got into difficulty in storm force winds off the south-west coast. The Shannon Sikorsky dropped two pumps on board the British-registered Invention after it reported it was taking water 120 miles south of Fastnet Rock, while RAF helicopters winched the ten crew of the Mar de Los Sargazos to safety after the vessel lost power 135 miles south of Fastnet.

Sadly, however, there were to be several more fatalities at sea before the month was out. A Portuguese fisherman was swept overboard a British-registered Spanish vessel 160 miles west of Mayo. The Shannon Sikorsky searched the area with a fixed-wing aircraft from Scotland — to no avail. Another member of the same vessel, the Slebech, died in Limerick Regional Hospital from injuries to his lungs and ribs, sustained while on the vessel on 12 November. And a 20-year-old Cork fisherman, Aidan Burke, was declared missing and presumed drowned after he was swept overboard the Sceptre, fishing out of Union Hall on 27 November. Mr Burke had been fishing for a short time and had only joined the vessel a fortnight before. An air and sea search located the boots he had been wearing. Visibility was so poor that the Shannon Sikorksy had to return to base.

The deaths as Christmas approached marked out 2000 as the worst year in a decade for fishing industry accidents, with 29 fishermen lost by the end of November.

17

BLACKSOD TO BENBECULA
— CROSSING BORDERS

Rescue pilots flying at night often refer to the 'fishbowl effect'. A vessel's engine has packed up in a storm, it is adrift and pitching helplessly in heavy seas at 35 to 40 degree angles. In misty conditions with no ambient light, all the pilots can see is this illuminated deck rolling upwards.

As Flight Lieutenant Ian Saunders of the Royal Air Force (RAF) explains, picture yourself inside a 'grey bubble' with no visual clues to tell you which way is up or down. 'A helicopter is normally hovered by the pilot using visual references to keep the aircraft either over a spot on the ground or alongside a vessel,' he says. In mist and darkness, and in a storm, the only visual references may be the lights of a boat and the wave flumes and spray — all of which will probably be moving around dramatically.

'Using a three-dimensional control you, the pilot, have to try and remain stationary alongside these lights. If you resolve that problem, you then have to move over the top of the vessel, where you can no longer see the lights, and remain stationary in that position whilst you winch.' The only way around this is to fly with sole reference to the instruments inside the cockpit, 'that is, the pilot must not look at the boat or the sea'. As always, the winch operator will draw the word pictures while working with the winchman and ensuring that the helicopter isn't going to hit any part of the vessel below.

This was the situation facing Saunders as captain of an RAF Sea King from Chivenor, north Devon, about 140 miles south of Mizen Head in the early hours of 30 November 2000. The British-registered Spanish fishing vessel, the *Zorrozaurre*, with a crew of 13 issued a Mayday at 2.30 a.m. Greenwich Mean Time and said it was sinking. The message was picked up by Valentia Coast Guard in Kerry and relayed to Falmouth Coastguard, as it was in the British search and rescue zone. The RAF Chivenor Sea King was scrambled by RAF Kinloss.

The flight would test the helicopter to its limits, as 140 miles south of Mizen Head was also 185 miles west-south-west of the Scilly Isles. Cork and Castletownbere to the west were among the diversion options discussed by Saunders with his crew of Rescue 169 — co-pilot Flight Lieutenant Richard Michael, winch operator Flight Sergeant Andy Batchelor and winchman Petty Officer Air Crewman (POACMN) Alan 'Arfur' Mullins, who was on exchange from the Royal Navy for a period with A Flight, 22 Squadron, RAF.

The co-pilot and winch operator had a quick chat about offloading rescue equipment to ensure they had maximum lifting capability. They thought about taking two pumps, and then decided to request support, and pump transport, from a Royal Navy Sea King at Culdrose. The Sea King was airborne at 3.45 a.m. and arrived at St Mary's to refuel at 5 a.m. Visibility *en route* was reduced to 500 metres in heavy rain while wind speed was south-east at 60 knots.

A Nimrod with call sign Rescue 51 was scrambled from RAF Kinloss to provide top cover, but technical problems on board the fixed wing made communications pretty difficult. However, the Nimrod did manage to inform the RAF pilots that the Royal Navy helicopter was going to remain 'on readiness' at Culdrose.

The Nimrod tried making contact with the *Zorrozaurre*, but there was no English speaker on board. A sister ship, the *Tpocachi Star*, picked up the signal and offered to provide translation, while several other vessels in the area said they would stand by to lend assistance. The pilots were able to establish that the *Zorrozaurre* was still taking water, but the influx had slackened and it would probably stay afloat till about 7.40 a.m. That would give the helicopter crew just 35 minutes if it had enough fuel; it estimated that it would reach the vessel by 7.05 a.m.

One of the fishing vessels which had offered to help came alongside the *Zorrozaurre*, but the seas were too rough to attempt an evacuation. When Saunders caught his first sight of the vessel — at 7.05 a.m. as estimated — it was 'dead in the water', to quote his own mission report. It was lying east/west and listing at about 30 degrees to starboard. 'A liferaft had been deployed but appeared trapped under the starboard beam amidships', he wrote afterwards. 'All 13 survivors were positioned in the lee of the bridge, also amidships. The position of the dinghy made it impossible to abandon ship and get clear, as the wind was blowing the vessel on to it . . . The crew were mindful that we probably only had 35 minutes before the trawler sank.'[1]

The Sea King hovered off the stern to check out the situation. The deck was cluttered with masts, aerials, crossbeams, girders, rigging, wires and equipment, and the only viable winching area was a two metre by one metre patch on the stern. The crew counted the waves to determine when there was a 'lull'. Winchman Mullins volunteered to go out on the wire, and to be ready for that break when it came. He would then try to reach the deck guided by his winch operator.

The crew agreed to try it. However, Mullins was out on the wire when the vessel pitched and rolled and he thumped into a crossbeam on the deck. 'Up, up, up, move back quickly,' Batchelor, the winch operator, roared. The pilot responded and Mullins was recovered on board. Fortunately he hadn't been injured, but this was due to the winch operator's quick reaction, Saunders said afterwards.

The winchman's hoist cable had come 'uncomfortably close' to a heavy duty rigging cable on the vessel during those few seconds. The winchman could have swung under the rigging wire, snagging his cable with 'potentially dire consequences', Batchelor noted afterwards.[2] Mullins offered to try again, but the captain decided it was too risky. They would try to establish a hi-line on the vessel and winch the crew up from the ship without the winchman's assistance down below.

Hovering overhead was proving to be extremely difficult, as was the task of trying to lower the hi-line. There were five attempts, during which three hi-lines snapped and five weights were lost. The winch operator considered joining two hi-lines together but knew this could increase the 'snagging hazards'.

The crew were wearing lifejackets but had no waterproofs or survival suits. They were huddled together in their jumpers, jeans, tracksuits and trainers, amidships of the starboard side of the wheelhouse, sheltering from the worst of the weather. Thankfully, as the pilots watched their fuel gauge, a hi-line was finally secured.

The winch crew knew their job would have been easier if the men were astern, but this part of the deck was being pounded by large waves and there wasn't much time to argue. The terrified fishermen weren't familiar with using the strops; a radio message had to be relayed to them explaining how they should be worn. All 13 were lifted on to the helicopter; several smelt of alcohol. The Sea King left the scene at 8.05 a.m. with the vessel still afloat below. It had enough fuel to make it back to the Scilly Isles. As it headed north-east, the winch crew passed hot drinks around the cabin to the shivering survivors.

It was 11 a.m. when the helicopter arrived at the Royal Naval base in Culdrose, and the crew were asked to attend a press conference to furnish details of the rescue. 'This was a highly demanding rescue, carried out in atrocious weather conditions and a long way from fuel, which constrained their time on the scene', the squadron's flight commander, Squadron Leader W. H. J. Webber, noted in the mission report. The captain and radio/winch operator had displayed 'great skill, courage and presence of mind', and the captain had made a number of 'sound decisions' that led to the rescue of the casualties and the safe return of his crew. He also paid tribute to the courage of the winchman in offering to be sent out on the cable a second time, in spite of the danger he had faced on the first occasion. The crew had been true professionals —all the more so when it emerged they had had only three to four hours' rest before setting out from Chivenor. They had been on search and rescue standby duty the previous day.[3]

Over a year later the RAF crew were informed they had been nominated for, and had won, the RAF Escaping Society trophy for 2001. Three of the crew were still at Chivenor at this stage, while Alan Mullins had been transferred to Yeovilton. The RAF Escaping Society was formed in 1946 by members of the British forces who evaded capture, or escaped detention, during the Second World War; they wanted to pay tribute to over 14,000 people who had risked their lives to help them.

To perpetuate the memory of those thousands — many of whom died for their efforts — a trophy was presented by the society to the RAF. It was intended to be presented annually for 'the best individual or collective feat of combat survival, or comparable feat of challenge and achievement carried out by personnel of the RAF during operations or recognised training'. Although the society closed in 1995 the trophy is still awarded. Previous recipients included a military liaison team in Kuwait and British military forces in Kosovo. It was awarded for marine search and rescue efforts in 1981 (RAF Valley) and in 1989 (to Flight Sergeant Dodsworth at the search and rescue wing in Brawdy).

There was a strange epilogue to the plight of the *Zorrozaurre*. It was still afloat when the RAF Sea King left it, carrying its 13 very relieved crew to shore. It was still afloat several days later, in spite of the estimate at the time that it would be under water before 8 a.m.

on 30 November 2000. Navigational warnings were issued, giving the last known location of the ship. Ten days later, the Naval Service patrol ship LE *Orla* was asked to follow up a report that a slight diesel slick and fish boxes had been seen in Bantry Bay, just east of Castletownbere in west Cork. It found a semi-submerged wreck, and Naval Service divers confirmed that it was the *Zorrozaurre*. It had drifted 140 miles north-east and had come into the bay 'like a submarine', Gene O'Sullivan, divisional controller of Valentia Coast Guard, said. It was a miracle it had not collided with anything in its path. 'Like a ghost ship,' O'Sullivan said.[4]

And a ghost ship it turned out to be, in more senses than one, for the Naval Service. The wreck had finally reached 'land' just 400 metres from its sister vessel, the *Nuestra Senora de Gardotza*. This was the vessel that had got into trouble in a storm in Bantry Bay just over a decade before and ran up on rocks east of Bere Island. Its crew were saved, but not without the loss of one of the rescue personnel — Naval Service Leading Seaman Michael Quinn.

BRITISH RESCUE SERVICES WERE back out again, assisting another British-registered Spanish vessel off the Irish coast on New Year's Day 2001. The *Pembroke* reported it had lost its rudder and was taking in water about 120 miles south of Mizen Head. A 13-strong crew of Spanish, Portuguese and African nationality were winched off the vessel by a Royal Navy Sea King from the Culdrose base near Helston in Cornwall and were flown to Cork Airport. A second Royal Navy helicopter was scrambled to assist, and both aircraft refuelled on the Scilly Isles *en route*. An RAF Nimrod from Kinloss was also dispatched to provide top cover. The survivors were taken to Cork University Hospital but had no injuries, according to the Irish Coast Guard. In the words of a local agent, they were 'just waiting now for their air tickets home'.[5]

ON 5 MARCH 2001 Falmouth Coastguard in Britain reported a 'hit' from an emergency position-indicating radio beacon (EPIRB). It contacted Clyde Coastguard in Scotland. The signal was coming from a 30 metre German-registered side trawler named the *Hansa* in Clyde's rescue area off the Scottish coast. Unless the vessel had lost its EPIRB overboard by accident, it was in difficulty about 240 miles west of Tiree.

Earlier that evening a very faint Mayday signal had been detected, which had produced no response when the Coastguard tried to return the call. Malin Coast Guard in Donegal had also picked it up and they too had been unable to make contact. When the EPIRB went off, a Canadian Aurora aircraft exercising in the area was asked to take a look. It was joined in the early hours of 6 March by an RAF Nimrod aircraft from Kinloss. A British Coastguard helicopter from Stornoway was scrambled, as was a Royal Navy helicopter from Prestwick.

As dawn approached, the Canadian aircraft spotted a flare and a strobe light, and the RAF Nimrod reported seeing two liferafts in the water. About 5 a.m. on 6 March the call came through to the Irish Coast Guard helicopter at Shannon. The crew were told there was a fishing vessel in trouble about 170 nautical miles north-west of Blacksod, Co. Mayo. They were asked to stand by. At this stage nine men from a crew of 16 had been picked up by the British Coastguard's Stornoway-based Sikorsky S-61 and had been flown in headwinds to Benbecula on the Scottish western isles.

A complex search for the seven missing crew was now under way involving the RAF, the Royal Navy, the Air Corps Casa maritime patrol aircraft, the British Coastguard, several NATO vessels on exercise, a British tanker, three Norwegian cargo ships and a number of Danish fishing vessels. The weather was deteriorating, with winds of 50 knots and 10 metre/30 foot seas. The Shannon Sikorsky was airborne in 30 minutes and flew in the dark to Blacksod lighthouse some 100 miles north to refuel. On board were Captain David Courtney, co-pilot Captain Chris Pile, winch operator John Manning and winchman Eamonn Burns.

Vincent Sweeney, Blacksod lighthouse's attendant keeper, was well used to visits from the Sikorsky and knew most of the crew on first name terms. The helicopter crew were told to stand by at Blacksod, so he made them a full Irish breakfast and plied them with cups of tea as they waited for further instructions. In the meantime they tried to get some weather information from Scotland to see if they would be able to land there after the mission. If they were tasked, it could be at the very edge of their limit and range.

There was no further word by 10 a.m. At 10.40 a.m. the pilots prepared to head back to Shannon for a crew change. They were 15 minutes into the return flight when they were asked to reroute to the rescue area. A survivor had been spotted in the water by the RAF Nimrod's master air electronics operator, Mark Lister. Fortunately he was wearing a dayglo orange survival suit. Flight Lieutenant Steve James said the waves were 'the size of a house' at 15 to 20 feet. The man would have had to contend with bitterly cold seas and 45 knot winds breaking on his face. The Nimrod dropped liferafts and made contact with a fishing vessel near by. The air crew relayed the man's latitude/longitude so that the vessel could pick him up.

Meanwhile the Shannon helicopter had calculated that the man was 17 miles further north than his ship's original position. The rescue site was now 188 nautical miles away, barely within the helicopter's range. Courtney landed again at Blacksod, took on the maximum allowable fuel at 4,750 lbs and took off again at 11.12 a.m. He and his crew estimated that even with this full load they would have just enough for ten minutes on the site but not sufficient to return to Ireland.

The further north they went, the more committed they were to a landing in Scotland after the rescue. However, they had no weather data for Scotland due to a computer problem in the meteorological office. 'We knew we were the survivor's only chance at the time, as the Stornoway S-61 had to turn back from its second attempt to reach the scene due to headwinds,' Courtney says. 'The Royal Navy at Prestwick had also been unable to get there. At around midday we at last obtained a weather brief for Scotland by satellite phone from the Air Corps Casa fishery patrol plane. The conditions were clear. We did our maths then — knowing we could land in Scotland extended our time on site to 30 minutes.'

The Shannon helicopter was approaching the rescue area at 12.30 when it learned that the survivor had been taken on board a fishing vessel. However, he had been over 12 hours in the water and would still require immediate medical evacuation. 'We searched the liferaft area for additional survivors and found none,' Courtney says.

'We routed to the fishing vessel to winch the survivor on board at approximately 12.50, with time running out. We opted for a hi-line rescue as the vessel was pitching and rolling dangerously. But after winchman Eamonn Burns was put on board, the hi-line snapped. With no time to deploy a second line, we repositioned for a standard lift.'

'We all know how many minutes we have, because the fact is that the calculations are done by all four of us,' Eamonn Burns says. 'So we all know exactly what we are doing. Because they observe, the backseat crew often know more — with all due respects!'

Burns remembers that the man was severely hypothermic. 'His brother had died while he was holding him. We got him into a stretcher, winched him off and made our way to Benbecula, escorted by the Canadian Aurora aircraft. The red lights [for fuel] were flashing as we landed.'

Even as the seconds passed, the captain had to obtain diplomatic clearance for a possible emergency landing *en route*, because NATO was exercising in the area. 'Thankfully the winds remained steady on our return and we landed at Benbecula at 3.24 p.m. with just 200 lbs of fuel. Phew!' Courtney exclaims. 'We returned to Shannon the next day at approximately 1.12 p.m. It was the furthest north the Shannon S-61 had ever flown at the time.'

The tenth survivor's ordeal was regarded as extraordinary. To last for almost 13 hours in the Atlantic in those conditions at that time of year is stretching the limits of human endurance. 'He owed his life to his survival suit,' one of the RAF Nimrod crew remarked. Unfortunately none of the six missing was found.

The Shannon helicopter crew subsequently received a marine merit award. Courtney emphasises that the Scottish Coastguard's airlift of nine crewmen earlier was far more challenging. Before leaving Benbecula, both crews stood for a photograph to mark a rare joint effort. 'They had to do that rescue at 4 a.m. — in the dark. We had daylight,' Courtney says.

'There is no comparison between night and day rescue. By day you can see much more. At night all efforts to hover the helicopter over a moving sea or a fishing vessel are made more difficult. Yes, the procedures are the same, but there needs to be a slowness that is almost zen-like to succeed. However, this must be balanced by the urgency of carrying out a rescue when lives are at stake. We are trained in underwater escape procedures. But if the helicopter has a catastrophic setback such as an engine failure, ditching at night is altogether more life-threatening.'

The limits of the Stornoway Sikorsky's endurance was tested on that occasion, but in November 2002 it carried out one of the longest civilian helicopter rescue missions on record in Britain. On 25 November 2002 it flew nearly 600 miles to lift an injured 22-year-old crewman from Sierra Leone from a Spanish fishing vessel. Alberto Okori skinned his hand while working on machinery when his vessel, the *Nuska*, was fishing 287 miles north-west of Stornoway.

The aircraft which was requested by Medico Madrid, the Spanish medical rescue radio service, had to refuel once on a drilling rig *en route* to the fishing vessel. The aircrew had just five minutes, in darkness, to carry out the rescue operation, even with the extra fuel on board. An RAF Nimrod provided top cover, and Okori was flown to the Western Isles Hospital in Stornoway. 'It was 50 miles outside our range, and the crew couldn't afford to make any mistakes with fuel so low,' the deputy chief pilot at Stornoway, John Bentley, said. In the 15 years that a rescue helicopter had been based at Stornoway, it was the furthest one had flown to rescue a casualty.

Just a month before the *Hansa* rescue in March 2001, the Air Corps search and rescue unit had been involved in another cross-border rescue, though somewhat closer to hand. At 6.25 p.m. on 19 February 2001 Malin Coast Guard received a call from Belfast Coastguard seeking helicopter assistance; the North had no night-time air-sea rescue capability. Fortunately the Finner-based Dauphin was already airborne when it got the 'shout', as it was on a training exercise off St John's Point in Donegal Bay.

A surfer, who had been out with ten others, had got into difficulties at Dunluce, north Antrim. His colleagues had been picked up by local fishermen and the Portrush lifeboat, and he had managed to climb on to rocks at the base of a 33 metre/100 foot cliff. Fortunately all were wearing wetsuits; apparently they had been filming attempts at 'rockhopping'. There was some suggestion that alcohol was involved!

The Portrush lifeboat made several attempts to reach the lone man on the rock, but couldn't because of the location and the heavy seas. It was far too dangerous for him to move. 'They knew it was going to be risky for us and I think they may have hesitated in calling us, but it had become clear that this was going to be the only option if he was to have any chance of surviving,' Captain Shane Bonner, pilot on the Dauphin that night, recalls. 'This isn't one we train for.'

Bonner (now Commandant), his co-pilot Lieutenant Anne Brogan, the first female Air Corps pilot, winch operator Sergeant Des Murray and winchman Corporal Aidan Thompson gleaned as much information as they could from the local coastguard unit on VHF channel 16. The coastguard turned on the headlights on their emergency vehicle to guide the Dauphin in. The Portrush lifeboat also stood off the cliffs and illuminated the area.

'With that sort of briefing, we knew then the only thing to do was to go straight into it,' Bonner says. Thompson remembers they did a 'high recce' or reconnaissance and then hovered over the sea. 'They put me down on to the rock, and it was rough enough. There was a bit of climbing and clambering, but I got to him. He was shivering with the cold, but I'd say he was more embarrassed than scared.' If the students had known anything about tides, they wouldn't have been caught out.

The helicopter landed at a car park beside a waiting ambulance and handed the man over. 'He wasn't in bad shape and wasn't hypothermic because of his wetsuit,' Bonner recalls. The helicopter crew then flew to Derry Airport for fuel and returned to Finner, knowing they had just completed a job well done. The crew were given a marine merit award by the Minister for the Marine, Dermot Ahern, in November 2002.

TASKINGS ACROSS THE BORDER were rare enough in the period before the introduction of the State's medium-lift service. The wheel was to turn full circle, however. When the RAF phased out its Wessex helicopters in March 2002, this included the Wessex search and rescue helicopter at Aldergrove. At time of writing, air-sea cover for the north-east is now provided by the Irish Coast Guard and the British Coastguard at Stornoway.

That night rescue at Portrush in February 2001 was not the first time Captain Bonner had been involved in cross-border missions. On Friday, 13 December 1996, a search and rescue Alouette from Baldonnel was scrambled. A helicopter which had been due in south County Down the previous evening had gone missing.

The SK76 helicopter owned by Norbrook Laboratories, an agricultural pharmaceutical company headed by Senator Edward Haughey, had left Belfast International Airport at Aldergrove after 6 p.m. on Thursday, 12 December, for Senator Haughey's home near Kilkeel, Co. Down. The last contact with the craft was ten minutes after take-off. The pilots were due to stay in Kilkeel overnight, so the helicopter was reported missing only when it had not returned to Aldergrove the following morning.

The RUC, Gárda Siochána, British and Irish armies, the RAF, the Air Corps and coastguard services mounted a land, sea and air search. Bonner remembers that the Alouette was asked to search Dundalk Bay. With him were Corporal Alan Gallagher and Airman Jim O'Neill. 'We did this for an hour and a half but found nothing. We routed to the British Army heliport at Bessbrook, just outside Newry, to refuel, and then resumed our search. The Coast Guard then asked us to search the mountains around the Cooley Peninsula,' says Bonner.

'We were up over the Cooley Mountains when O'Neill, who was on the winch, commented on what looked like a rubbish tip high up on one of the mountains. From a distance it looked like scattered paper along the slope. I got a bit of a chill down my spine, because

I knew no one would go that high up a mountain to dump rubbish. As we got closer we made the unfortunate discovery that it was indeed the crashed helicopter. We lowered Jim near the site to look for survivors, although it was immediately obvious to us that this was not a survivable crash.'

The discovery of the SK76 wreckage was made at 3.20 p.m. at Raven's Rock on Carlingford Mountain, west of Omeath. O'Neill found two bodies in the wreckage and a third near by. The Alouette notified the authorities, and the gardaí and mountain rescue were dispatched to recover the bodies.

The three pilots were named later as Kevin Mulhern from Oxfordshire, Jeremy Wright from Cheshire and John Smith from Norfolk. The SK76 appeared to have strayed about three miles off course in the Carlingford Lough area. At the subsequent inquest it was stated that the three pilots had died from multiple injuries.

Bonner found himself back in Portrush, Co. Antrim, in late May 2003 when one of the Dauphins at Finner was involved in a search and rescue demonstration with the Portrush RNLI. The demonstration on 24 May had been a success and the Dauphin was on its way back to Finner when it picked up a 999 call on channel 16 from Belfast Coastguard. A swimmer was reported to be in difficulty close to Portstewart.

Belfast Coastguard asked Portrush lifeboat to investigate it, but the Dauphin was over Portstewart at the time and said it would check it out. On board with Bonner were Lieutenant Tom Kelly, Sergeant Colm Blackburn and Corporal Ciaran Smyth. 'As we approached Portstewart Harbour, I noticed some people standing on nearby cliffs waving at us and pointing into a cove beside the harbour. We saw a man in his early twenties in the water in this cove, which was horseshoe shaped and surrounded on three sides by jagged cliffs,' Bonner says.

'The casualty was in the water right at the foot of the cove, and some people on top of the cliff had thrown him a rope which he was clinging to. The sea inside the cove was extremely rough, due to the funnel effect. It was entirely "white water" and the level at the base of the cliff was rising and falling by as much as 20 feet or so. This meant that the man could not climb out of the water using the rope he had been thrown. Any time he tried, the sea washed him back down again.

'As soon as we got the casualty in sight, the winch operator remarked that the man was in big trouble and that if we didn't get him soon, he would be lost. We immediately prepared to winch. I lowered the helicopter into a winching position while the rear crew prepared their equipment. I have always had nothing but the greatest of respect for the winchmen who work for us, and for their bravery and commitment to saving lives, and this rescue only strengthened that respect,' continues Bonner.

'We lowered Corporal Smyth into this huge swell and he was immediately dashed against the jagged rocks along with the casualty. I managed to maintain the helicopter in a hover while he attempted to get the casualty into the rescue strop. As you can imagine, this wasn't an easy task as both men were being constantly buffeted and submerged by the incoming waves. Meanwhile the co-pilot was helping me keep the rotors clear by advising me of the blades' proximity to the cliff on the left, and the winch operator was advising me of the proximity of our tail to the wall behind us, as well as monitoring the winchman's progress. After a minute or so the winchman had the casualty secured and we hoisted both of them out of the water.'

Ciaran Smyth on the winch says the sea state was 15 to 20 feet. 'The swimmer was only wearing a pair of shorts, and he was very weak when we got to him. I'd say he had only a minute or so left, really. I was pushed up against the cliff and had to try and get him into the strop, and I swallowed a lot of sea water while I was doing that! I got a couple of bruises, yeah, and my helmet was quite badly chipped when I looked at it afterwards. Normally we don't wear a helmet on a wet lift because the saltwater corrodes it, but luckily this time I did. I remember I told him to let go of the rope, but he didn't — partly I suppose because of the way it was tied around his hands. So we lifted him up with the rope and someone cut the other end. He had a couple of lacerations to his head and was really in quite a bad way. Although he was conscious in the water, he was drifting a bit on the way up and I had to give him a couple of slaps in the face to try and keep him awake. So I'd say anyone watching on the cliff top got a good view of me clattering this poor fella, and wondering what I was at at all!' Smyth says.

The pilots routed to Altnagelvin Hospital in Derry. 'On the way to the hospital the rear crew administered first aid because the man was hypothermic, badly battered and extremely weak. He also had a bad gash on his forehead which was bleeding. The crew had to work hard to keep him conscious,' Bonner says.

'I firmly believe that if he had spent a further minute or two in the water, he would not have survived,' Bonner notes. He also doubts that the lifeboat would have been able to reach him, due to the configuration of the cove and the strength of the waves against the cliff. 'And whoever threw him the rope acted very wisely indeed. Without this lifeline to cling to, the man would almost certainly have died by being dashed against the jagged rocks.'

Bonner concluded: 'It took 30 seconds to reach the casualty from the moment the 999 call was made. We could as easily have been carrying out a demonstration in Galway or even at Finner. Either way, the only way this man survived at all was due to the fact that we just happened to be in that area on that Saturday afternoon. It really was a case of being in the right place at the right time.'

18

THE *CELESTIAL* DAWN

The 'deserted' Inishkea Islands off north Mayo's Mullet Peninsula are renowned for their populations of barnacle geese and grey seals, and were once home to pirates and poitín-makers and a Norwegian whaling station. They also have a rich archaeological heritage; but the settlement which quit in 1927 after 45 west coast fishermen were lost in the Cleggan Bay disaster dated only from the 1700s.

On a good day there's a sense of undisturbed paradise about the uninhabited outcrops. On 27 July 2001 the cargo of sheep on Tony Lavelle's boat were probably too terrified to think about paradise — or the feed of grass before them. Mr Lavelle, former owner of the Mullet Bar in Belmullet, was transporting the animals from Frenchport to Inishkea north with his 14-year-old son Anthony. Conditions seemed good as they set out on the five mile sea journey.

It was just after midday, 12.29 p.m. to be precise, when a local fisherman noticed something amiss. He spotted some sheep on Duffer rock, about halfway between the mainland and the Inishkeas. He also saw some sheep's carcasses floating in the water and a small slick of oil close by. He raised the alarm, and the Dauphin helicopter at Finner camp and Ballyglass lifeboat were called out. The Naval Service patrol ship LE *Ciara* was also in the area and offered its assistance, along with several local boats.

At 1.14 p.m. the fishing vessel, the *St Theresa*, came upon a man hanging over a timber oar and took him on board. He was unconscious. The Finner helicopter was just arriving on the scene, flown by Captain Donagh Verling and co-pilot Lieutenant Anne Brogan. 'We had been trying to get information *en route*,' Brogan remembers. 'There were so many sheep in the water below us that it made searching very difficult, even though visibility was fine. When we took Mr Lavelle into the helicopter, he was hypothermic. He was trying to tell us something, but he just couldn't speak.'

The helicopter flew him immediately to Mayo General Hospital in Castlebar, where he was rushed to intensive care and described as 'critical but stable' that night. It was the only decision to make: the doctors told the Air Corps crew that if he had been another minute in the water, he wouldn't have survived. 'We flew back out to look again,' Brogan remembers.

'Our adrenalin was up. We felt we had saved someone's life. Then we heard from Shannon that there had been a 14-year-old boy with him, and that both of them had been clinging to wreckage. We had been on such a high, and then our hearts literally sank.'

The *Ciara* under Lieutenant Commander Martin McGrath co-ordinated the search, assisted by the Ballyglass lifeboat. The Sikorsky helicopter from Shannon took over from the Air Corps Dauphin, and up to 30 smaller vessels including a North-Western Regional Fisheries Board patrol vessel continued their efforts to find the young boy. The boat was found in about 30 feet of water close to Duffer rock, over a mile from where Mr Lavelle was rescued. Local divers searched the wreck and found the bodies of several sheep trapped in the forward area. McGrath noted that a northerly drift had probably carried father and son a considerable distance and may have hampered their efforts to swim for shore. Neither had been wearing a lifejacket.

Thick fog enveloped the area shortly after 8 p.m. and the search was stood down at 8.30 that night. The plan was to resume at first light. Mr Lavelle had only recently sold his pub licence and bought the boat. The tragedy happened just eight days after two young Belmullet children, Tish (15) and Niall Murphy (5), drowned off the Mullet Peninsula. Niall had been playing with an inflatable toy which the family had only recently purchased, and was caught by the tide. His sister had tried to rescue him, only to get into difficulties herself about 100 yards from the shore.

Brogan remembers that the aircrew were 'pretty subdued' by the time they returned to Finner. The Naval Service patrol ship was referring to 'tango four' — in other words, the young boy was presumed dead. 'We spent hours talking about it afterwards that night. You always hope things will work out; that was one occasion that it just didn't.' It emerged that Anthony Lavelle, a pupil at St Brendan's College in Belmullet, was a strong swimmer who had done his best to keep his father, a non-swimmer, afloat. He was trying to cling to a five-gallon plastic drum which was too buoyant and kept rolling over. Exhausted, he lost his own fight to survive only minutes before his father was recovered. 'That's what Mr Lavelle had been trying to tell us as we flew away,' Brogan says. To date, Anthony Lavelle's body hasn't been found.

Lieutenant Brogan, from Dublin and of County Galway stock, was the first female helicopter pilot to earn her wings in the Air Corps. She was studying aeronautical engineering at Strathclyde University in Glasgow, Scotland, when she was notified that she had been accepted as a cadet in the Air Corps. After four years on search and rescue from 1988 to 2002, she switched to flying solo with the Garda Air Support Unit — the aim being to acquire enough hours to become a captain back on rescue missions. 'Search and rescue is an immensely satisfying job,' she says, and she wouldn't want to do anything else.

ON NEW YEAR'S DAY 2002 an Air Corps crew were just finishing lunch at the daytime search and rescue base in Waterford Airport when there was a scramble call from the Irish Coast Guard. A swimmer was in difficulty in Rinnishark harbour off the 'large Burrow' or sand dune in Tramore Bay; a friend who had also been in the water had made it to shore and had alerted some walkers on the beach.

The crew changed into immersion gear, and it took just four minutes to get from Waterford Airport to the bay. The pilot, Lieutenant Lee Brennan, could see several people at the water's edge on the bay's eastern shore and they were pointing west. Initially the winch operator Corporal Derek Everitt and winchman Corporal Davitt Ward couldn't see anything, but then they spotted a man directly in front of the helicopter, swimming 'sluggishly' towards the shore which was 50 to 60 metres away. He wasn't making any progress, due to the outgoing tide and the current flowing from the estuary linking the Black Strand lagoon to Tramore Bay. It wasn't far from where the Dauphin had crashed with four aircrew on board in July 1999.

The pilot flew a very tight circuit towards the swimmer, and Everitt set up for a wet lift while keeping an eye on the casualty in the water. He 'pattered' the pilot into a hover position just over the man, and Ward was winched down immediately on the cable. The swimmer, a teenager, appeared to have seen the helicopter. As Davitt reached his hand out to him, he rolled on his back and began to slip under. This was one of those classic situations where casualties can give up the struggle when help is at hand.

Davitt managed to pass the strop over his arm and around his back till it was under his chin. The young man was in the early stages of hypothermia; he would require two strops to allow for a horizontal lift, but Davitt was doing his best to keep a grip on him and keep his head above the surface. The horizontal lift technique had been developed by the RAF after the Fastnet yacht disaster of 1979; one of the recommendations in the aftermath of the subsequent rescue operation was that horizontal lifting could diminish the worst effects of hypothermia.

The winch operator and pilot agreed to lift the two men to the door without using a second strop. Everitt managed to wind them up, but the trouble was only starting. The swimmer was barely conscious, wearing only togs, and couldn't help himself in. What's more, he was larger than either of the crewmen, weighing about 180 to 200 lbs. The winch crew struggled to keep a hold of the 'cold and slippery' casualty, but he was in danger of falling out of the strop and he wasn't in the aircraft yet. It had resonances of the situation in which Commandant Tom O'Connor and his crew found themselves off the Forty Foot in Dun Laoghaire in March 1993.

Brennan could see what was going on over his shoulder and decided to 'hover taxi' the 70 metres to the beach and land, to allow the crew better leverage. At this stage Davitt was still out of the aircraft, facing in to the door with his legs jammed on the floor and keeping his knees under the man's crotch to prevent him from falling. 'He was facing towards Davitt,' Brennan remembers. 'If he'd been wearing clothes, there wouldn't have been a problem as we'd have had something to pull on to drag him in.'

Brennan landed the aircraft and the young man was hauled in to the rear seat, covered with a thermal blanket and given oxygen. The pilot took off again and flew the six miles to Ardkeen Hospital in Waterford. The young man was able to give the winch crew his name — Tristan John McGrath — but no other information. They described him later as 'incoherent, combative and unresponsive', the classic signs of hypothermia. As the Alouette approached the hospital, the pilot saw an ambulance driving through the gate. 'It was the last ambulance on the campus and it was leaving. There was only a skeleton staff on for New Year's Day.'

This presented the crew with another challenge. The pilot jumped out of the helicopter and ran towards the casualty department. 'I told the nurses I had a seriously ill man on board and I needed a doctor or an ambulance, or both. One of the nurses gave me some blankets and I just legged it back to the Alouette.'

By this time the winch crew had stripped off their immersion suits and were using their own body heat to keep the teenager warm. A Garda on security arrived and offered to help bring equipment down to the pad, but a stretcher was needed. Some 20 minutes later an ambulance arrived and McGrath was taken to casualty. 'Have we landed yet?' he asked, when he was being carried by stretcher out of the back of the vehicle.

The Alouette crew were back at Waterford when one of the doctors in the hospital phoned. Would they know if the young man had breathed in any water? The medical staff was having a problem trying to get any heat into him at this stage. His core temperature had been measured at 28 degrees Celsius. 'Most people give up and lose consciousness at 32 degrees,' Brennan noted. In his subsequent report to the Irish Coast Guard, Brennan raised the issue of the 20-minute delay at the hospital landing pad.

The following day the same crew were on duty at Waterford when they had unexpected visitors. It was Tristan John McGrath and his mother, and they had come to say thanks. He was wearing a wide-brimmed hat. They were members of the Amish community in Dunmore East, Co. Waterford.

It emerged that McGrath had been in the water for an hour when he got into difficulty, having decided to take a New Year's Day dip. His parents had warned him not to swim in that location. 'He had no recollection of the rescue at all, but was very grateful none the less,' Brennan says. 'His strength and fitness were such that he had recovered in that short period of time and complained only of a bad headache.'

The young man had a camera with him, and he and his mother took photographs together with the aircrew. Brennan calculated later that it had taken a total of nine minutes from the time the scramble phone rang at Waterford to the time where the casualty was secured and on board the aircraft. Total flying time had been 25 minutes. McGrath couldn't have got into trouble at a better location, but had there been any further delay at the hospital, events might have taken a tragic turn.

IT IS NOT OFTEN there are witnesses to an air-sea rescue, but there were quite a few watching winchman Peter Leonard on 2 February 2002 when an Irish-registered fishing vessel, the *Celestial Dawn*, ran aground at the foot of a cliff marking the eastern entrance to Dingle

Harbour in County Kerry. The vessel had been *en route* out of the harbour in bad weather when the alert was raised. From the shore, ten men could be seen clinging to the port rails with heavy seas breaking over them.

Valentia Coast Guard scrambled the Sikorsky S-61 at Shannon, the Valentia lifeboat and the Dingle Coast Guard unit, while local boats also responded to the alert. The wind was gale force seven, south-westerly, making an approach by any of those local boats impossible. By this time a large number of people had gathered on the headland, watching helplessly as the terrified crew clung desperately to the guard rails — trapped on the listing vessel, so close and yet so far from the shore.

'It wasn't the heaviest of seas, but there was a big one running,' Captain Derek Nequest, pilot of the Shannon Sikorsky tasked to assist, remembers. His co-pilot was Captain Mark Kelly, while John Manning was on the winch and Peter Leonard was winchman. 'The waves were smashing the wreck, lying on its starboard side and moving quite a lot. If any of those crewmen had lost their grip and gone into the water, there was no way that anyone could have saved them.'

A few of the crew had lifejackets, but most didn't. As the Sikorsky flew over, the local coast and cliff rescue unit illuminated the area, and the lifeboat was also beaming its light from the sea, and hover conditions were very good, Nequest says. As Leonard was lowered on the cable, there were a few expletives from John Manning on the winch as his winchman was caught by waves.

'I took up one at a time because of the conditions,' Leonard says, and recalls that one man refused to let go of the guard rail. 'Myself and one of his colleagues had to prise his hands off the rail and get the winch strop around him. One of his colleagues was shouting in Spanish, but he just didn't want to move.' The helicopter flew the traumatised crew to Kerry Airport.

Journalist Ted Creedon, one of the crowd of people on the headland who had been among the first to raise the alarm and who watched the helicopter at work, later wrote this account for *The Kerryman*.

> It was 7.20 p.m. on Saturday night when my brother-in-law called to say he had seen a red flare towards Dingle lighthouse, about a mile from our house. I rang the harbour master, Lieutenant Commander Brian Farrell, but he was already

heading for Beenbawn, a small beach near the lighthouse.

I drove there with my daughter Siún, and arrived just as another red flare burst over the headland near the lighthouse. The wind was strong and the waves thundered against the cliffs but there was nothing to be seen in the storm tossed seas around the harbour's mouth. Brian Farrell and his son Niall were in the car park, gearing up for the trek across the cliffs in the direction of the flares. Fisherman Tom Kennedy was there also, and it transpired later that he had been watching from his house as the *Celestial Dawn* left port. When her lights suddenly disappeared he raised the alarm.

Siún and I went home, got changed, borrowed a torch from a neighbour and set out across the cliffs to the lighthouse with Jimmy Bambury, Dingle fire brigade chief. The cliff path is hazardous in daylight but ten times more so in the dark.

The first thing that struck us was the stench of diesel fumes on the wind. Jimmy, who had information that a Spanish trawler was in trouble, said: 'I don't like that smell, lads. She must be gone down.' We arrived at the lighthouse and saw several people with torches pointed towards the sea below.

We didn't know what to expect as we looked over the cliff and for several heart-stopping moments it was difficult to comprehend the terrifying sight that met our eyes. Fifty feet below, the *Celestial Dawn* was lying half submerged on her starboard side, with her crew clinging to the railings all in a line along the side of the ship.

Most had lifejackets but some didn't and those who didn't were taking the full force of the waves as they broke over the trawler's stern. The vessel was being battered against the rocks on either side of her. Sometimes the swell lifted her high in the air and she bucked like some metal monster, then settled again for a few moments. When this happened the frightened men fought desperately to hold on to the railings. We wondered how long they could survive that.

The Dingle Coast Guard rescue boat was on the scene and came as close to the trawler as the cox, Mark Greeley, dared to go. Local ferryman Jimmy Flannery had his boat, the *Dingle Bay*, positioned so as to receive the fishermen if the Coast Guard dinghy could effect a rescue. The fishermen were shouting and gesticulating but the situation was too risky for the rescue boat. There was a real danger of her being caught in drifting nets or ropes or of being bashed against the trawler's hull.

There was also a language problem and verbal communication was almost impossible in those conditions anyway, but the rescue crew pointed to the night sky and shouted, 'Helicopter! Helicopter' at the fishermen to reassure them that help was on the way.

Local fishermen, watching from the cliff, feared that the men below might try to jump into the water to reach the rescue boat. That could have been fatal. Lieutenant Commander Farrell, who was the co-ordinator on the cliff, informed us that the rescue helicopter was *en route*. Valentia lifeboat arrived just after 8 p.m. and her very presence must have brought vital psychological relief and hope to the fishermen below as it did to those of us who were helpless observers on the cliff above.

There was a whoosh as she fired a flare over the scene. When it burst, an eerie pinkish light lit up the whole area from the lighthouse to the Towereen Bawn on the other side of the harbour's mouth. This allowed the lifeboat crew to appraise the whole situation. Then she switched on her powerful searchlights and trained them on the *Celestial Dawn*.

The trawler's liferafts floated between the vessel and the rocks. There was no escape that way, there was no access to the cliff and she was now sinking deeper into the sea. The men continued to cling to the bucking vessel as the waves broke over them. One man was seen to throw up; he may have swallowed sea water and diesel. We were asked to count the crew; and when it was certain that there were ten men on board we knew that none had been lost. The lifeboat and the local Coast Guard unit tried to get additional lifejackets to the stricken crew but encountered problems.

Then the sound of the Shannon-based Coast Guard helicopter was heard above the wind. She came from the east, made a long sweeping circle over the town and approached the harbour's mouth from the west. She positioned herself immediately over the *Celestial Dawn*, her searchlights trained on the men below. The powerful downward thrust of her rotors whipped up a storm of spray adding another visually dramatic element to the incredible scene.

After a few minutes, winchman Peter Leonard was lowered towards the trawler. We wondered if he'd make it or would he be battered against the vessel's hull in the wind and waves. We probably forgot for a moment that these rescue crews are highly trained and vastly experienced. Any doubts we had were immediately dispelled because the winchman had no sooner landed safely on the hull than he had securely harnessed one fishermen and both were hoisted to the Sikorsky.

We watched in admiration as he repeated the exercise and one after another the crewmen were lifted to safety while the helicopter remained rock steady in the buffeting winds over the trawler. The roar of her engines drowned out every other sound. I looked along the line of faces on either side of me. All were transfixed by the scene below. I noticed one young girl, perhaps a fisherman's daughter, with both her hands pressed palms together under her chin. The light from the rescue operation reflected on her face. She was praying silently.

One crewman seemed reluctant to leave. He may have been more terrified than the others or perhaps he was too cold to move. The winchman appeared to force the man's hands from the railings before lifting him clear of the vessel. The ten men were rescued in less than 15 minutes. Most of those watching had seen rescue exercises in the past but nothing like this. This was the real thing — a life and death drama being played out before our eyes. It may have been a routine exercise for the Coast Guard. To those of us watching from the cliffs it was a remarkable demonstration of training, skill, courage and experience. Many of those involved in the rescue were volunteers but all were focused and determined to save those ten fishermen.

As the helicopter rose and swung away with her precious cargo I felt like applauding, my faith in the rescue services fully affirmed. I told my daughter about the girl I had seen praying. 'I was praying too, Dad' she replied. We headed home across the cliffs as the lights of the Valentia lifeboat disappeared into the blackness of Dingle Bay.[1]

'There are very few rescues where you can say you have definitely saved ten lives,' Nequest admits. 'It would have been very difficult, if not impossible, to get them all off by boat in those conditions, and if there had been an attempt by boat a couple of lives might have been lost. That's one situation where the helicopter is ideal.'

The Sikorsky captain and crew were awarded a letter of appreciation for meritorious service at an award ceremony hosted by the Minister for the Marine later that year. Peter Leonard was conferred with the Michael Heffernan silver medal for marine gallantry for 'extreme courage' and his willingness to 'put his own life at risk to save others'.

19

HOURS OF BOREDOM, MOMENTS OF TERROR

The north-west Irish coastline is not the most hospitable at any time of year, but ships *en route* from Europe to North America in the winter months will avoid it if they can. The North Sea/English Channel route affords ships' masters the shelter of the British coastline before taking on the mighty Atlantic. However, in recent years competitive pressures have forced ships' masters to take the shorter, more hostile passages. Deadlines have to be met, whatever the weather — and the Panamanian-registered *Princess Eva* wasn't getting much of a break from the proverbial elements in late January 2003.

The Japanese-built single-hulled tanker was carrying 55,000 tonnes of gas oil from Copenhagen in Denmark to Corpus Christi in Texas when it hit a violent storm. Winds were so bad some 130 miles off the Irish west coast that the liferafts were at risk of being washed off. Three of the largely Argentinian crew, Pedro Sanchez, Andres Manrique, and Carlos Hernandez from Uruguay, ventured out to secure one of the rafts that had come loose. A large wave struck the hull, throwing the men across the deck. One of them was killed instantly and two sustained serious injury. Hernandez remembered that the waves were coming from the east and he was struck from behind. When he regained consciousness later in hospital, he learned that his leg had been amputated when he landed against deck machinery.

The Irish Coast Guard Sikorsky S-61 at Shannon received the call from Valentia Coast Guard, via Clyde Coastguard, at about 1.30 p.m. The helicopter was airborne at 1.42 p.m. The pilot was Captain Robert Goodbody, flying with co-pilot Captain Tony O'Mahony, winch operator Eamonn Burns and winchman Neville Murphy. 'We'd been told they were 130 miles north-west of Blacksod and there were three casualties, one fatal, one with head injuries and one with leg injuries,' O'Mahony says.

'We looked at the position, decided we could do it, and we headed to Blacksod to top up with fuel. We checked the position of the ship again, checked our fuel plan and got a weather update from the Irish Coast Guard by fax to Blacksod.' Winds were force 11, north-west — right on the helicopter's nose.

'We left there at 4 p.m. and moved out to the ship. An RAF Nimrod, Rescue 51 from Kinloss, was ahead of us to provide top cover. About halfway out we got word from the Coast Guard that a second man had died. When we got there it was 5.45 p.m., twilight. We'd been talking to the Nimrod, which had already taken a look at the *Princess Eva*. The ship was heading back in our direction at about ten knots.'

The helicopter flew in over the ship, but the position wasn't ideal for winching. There was a risk of severe turbulence over the superstructure, but the master of the vessel couldn't turn to try and reduce it because the storm was just too bad. Winchman Neville Murphy went out on the cable about 100 feet above the deck. 'It was higher than normal, but there was nothing else we could do. We didn't use the hi-line initially as we were directly over the deck,' he remembers.

'We used the hi-line then to get a stretcher down, and the crew on board had the casualty in a stretcher just inside the superstructure. They had given him drugs, and he was in and out of consciousness and very critical. He had lost a limb which was never recovered. I assessed him, gave him oxygen, bandaged him and made sure his vital signs were OK before we took him up,' Murphy says.

'The wind was causing a big bend in the wire, the seas were between 10 and 15 metres and the ship was pitching and rolling. There were lots of derricks on deck, so there was a real risk the wire would get snarled up, even though the ship was pretty big,' O'Mahony and Murphy explain. 'As it turned out, it was pretty straightforward. The fact that the ship's crew had prepared for it and had him in the right place was a big help, and we were away in 20 minutes and *en route* to Galway. On the way in, the casualty was coming and going, very disoriented and looking for his mother. His vital signs were lowering all the time, and he was very lucky to come out of it. He was as low as you can go when we got him into accident and emergency.'

The *Princess Eva* continued on course to Donegal; it had no refrigeration on board and would have to put the bodies of its two crewmen ashore. It anchored in McSwyne's Bay off Killybegs and was about to sail again with a relief crew when cracks were detected on the deck.

A detention order was issued by the Maritime Safety Directorate and anti-pollution measures were put in place by the Irish Coast Guard. Donegal Bay is an environmentally sensitive area, with several categories of designations and blue flag beaches. It would be 6 March before the vessel sailed again for Argentina — its cargo of fuel having been trans-shipped successfully on to two tankers under the supervision of Irish Coast Guard personnel.

The Irish Coast Guard's response, from the time the first alert was raised, received much praise from the prestigious maritime daily, *Lloyd's List*. The Irish had shown 'how it's done', the newspaper said, comparing the Coast Guard's handling of the incident to the Spanish Government's response when the single-hull tanker, the *Prestige*, ran into trouble off the Galician coast three months before on 13 November 2002. The Spanish authorities refused to allow the ship into port; it was leaking and badly damaged when towed out into the Atlantic and it sank six days later — causing major pollution on the Iberian and French coastlines.

Even worse, the unfortunate master of the *Prestige*, Captain Apostolos Mangouras, was arrested and detained for 85 days in a high security prison in Spain before being released on bail. All the master had done was try to save his ship and the lives of his crew, as Captain Kieran O'Higgins, president of the Irish Institute of Master Mariners, pointed out.[1]

THREE DAYS AFTER THE *Princess Eva* rescue, the Irish Coast Guard's marine emergency plan was activated in the Irish Sea when a 'pan pan' alert was issued by a ship about four miles south of Tuskar Rock. This was no ordinary ship, but the Stena Line passenger ferry, the *Stena Europe*, which was *en route* from Rosslare, Co. Wexford, to Fishguard in Wales with 220 people on board. Force eight northerly winds were gusting between 35 and 40 knots and the vessel was drifting at a rate of seven knots towards Carnsore Point.

Rescue helicopters from Dublin and Waterford were scrambled and assistance was sought from the RAF at Valley in Wales and Chivenor in north Devon. RNLI lifeboats at Arklow, Co. Wicklow, and Rosslare and Kilmore Quay, Co. Wexford, were launched, but by the time the Rosslare lifeboat had arrived on the scene, the ship's engines had been restarted. It had been a worrying hour, but the Irish Coast Guard was delighted at the speed with which its emergency plan had been executed.

Hoax calls are a reality of life for air-sea rescue crews, and there have been periodic spates of them off the west coast. There are also the taskings which turn out to be very different from the 'emergencies' first signalled — a fight on board a vessel, or a crewman who simply wants to get home, has led to abuses of the system. Incidents involving drug overdose at sea have become a trend, according to Irish Coast Guard crews. There are also several legal actions pending against rescue personnel at time of writing — reflecting an increasingly litigious society ashore.

Captain Derek Nequest of the Irish Coast Guard helicopter base at Shannon prefers to err on the side of caution when it comes to responding. 'You may get someone who is not as badly injured as you were led to believe when you were tasked; but from our point of view, we'd much rather be sent to the job and find out that he or she is not badly injured, than not to be sent at all and to find out afterwards that someone had suffered.'

Several months after the *Princess Eva* incident, the Shannon crew were involved in one tasking which proved to be very risky for the aircrew. A south-westerly storm lashed the Irish coastline in the middle of March 2003. Thousands of homes in west Limerick, Kerry and Cork lost electricity and a house was badly damaged in a lightning strike in Kerry. Fire brigade units were called out to deal with flooding on the roads.

In the midst of all this mayhem ashore, the Shannon helicopter was tasked in the early hours of 14 March 2003 to lift an injured crewman from a Spanish vessel some 40 miles south-west of Baltimore, Co. Cork. It was reported that the man had head injuries. He was one of 15 crew on board the 120 foot stern trawler, the *Nueva Confurco*, and he had been knocked unconscious when he was struck by a steel warp.

An emergency call was put out shortly after midnight, and the helicopter battled south-westerly gusts of force eight to nine to reach the vessel, arriving at the scene at 2.45 a.m. There was a 20 foot swell; putting a winchman down would be extremely difficult. The crew made several attempts. 'It was a very unusual sea,' winchman Neville Murphy explained afterwards.

'Normally you have a few waves coming in and you have a break of a couple of seconds —that's the critical point. We'd put the winchman out behind the deck, sneak in and wait for that calm patch. But on this particular night there was no break whatsoever, just a constant roll — the most unusual sea state I have come across. The ship just didn't settle, and the trawler's stern was packed with nets, net drums and clutter, which made winching more difficult. The bow is impossible in seas like that. The lads went into a hover, but found it hard to hold because of this large, long swell.'

The pilots decided to return to Cork to refuel, and a second attempt was made at 6 a.m. as the vessel headed for the coast. By this time the Irish Coast Guard had alerted the Castletownbere lifeboat and had asked the vessel to head for the west Cork fishing harbour.

The trawler was in Bantry Bay when the helicopter made its third attempt — a successful one this time, at about 8.30 a.m. They flew the injured man to Cork University Hospital, and heard about half an hour later that he was recovering well.

'As far as we were concerned, he was unconscious for most of the night, and you just can't take the chance,' Murphy, formerly of the Air Corps, says. He concurs with Nequest that it is always better to respond. He points out that during the early stages of the *Princess Eva* call-out, the aircrew had been told that one of the two injured crewmen had 'minor head injuries' and had been given two Panadol. The injured man died as the helicopter was *en route*. 'When I get down on the deck, I'll assess the patient anyway and make my decisions, so that's why we prefer to go,' Murphy says. 'It's nice to know the details on the way there, but it is not *need* to know.'

SEARCH AND RESCUE DUTY can involve a lot of sitting around — 'hours of boredom interrupted by moments of terror', as several winchmen put it. The Air Corps north-west base at Finner was in the throes of training up for the first medium-lift helicopter — a leased Sikorsky S-61 — in the spring of 2003, when it had one particularly active weekend which involved a rather harrowing experience for one of the winch crew.

Just after lunch on Sunday, 20 April, the scramble phone rang. It was Malin Coast Guard in Donegal reporting that a 38 foot stern trawler, the *Lady Christine*, was taking in water some 35 miles north-west of Sligo with three crew on board. The helicopter, flown by Commandant Donagh Verling and co-pilot Captain Dave Corcoran, arrived at 2 a.m. and the winchman Corporal Ciaran Smyth was lowered on deck by winch operator Sergeant Jim O'Neill. However, during attempts to lower a pump on board the vessel, two hi-lines from the helicopter snapped.

Corcoran said that at one point the situation was very hazardous with a 100 kilo pump swinging under the helicopter. The stern trawler was out of control, abeam to the wind with no engine power and at the mercy of the helicopter's down-draught. The Arranmore lifeboat and a fishing vessel, the *Silver King*, had come to assist and were standing by.

'At this point our fuel situation was critical, and so we left Corporal Smyth on board the vessel while we made a hot refuel in Sligo Airport,' Corcoran recalls. When the helicopter returned 30 minutes later the Arranmore lifeboat was attempting to move a pump across to the *Lady Christine*. 'They managed to get the pump operating and this stabilised the situation, but the vessel was still taking in water,' he says. The helicopter then tried to lower its pump with its remaining hi-line, but at this point the vessel's skipper decided to abandon the trawler.

The three crew and the Air Corps winchman were taken on board the Arranmore lifeboat, and the *Lady Christine* sank within seven minutes. 'We took Corporal Smyth up at this stage and the lifeboat took the three people to Killybegs,' Corcoran says. No injuries were sustained by the crew.

The rescue was the fourth involving the Finner helicopter over that particular weekend. On the previous Friday night it had been called out to a medical evacuation from Inishbofin Island. Then on Saturday afternoon it rescued a diver who had been swept north by currents north of Fanad Head, Co. Donegal. The diver had been carried two to three miles from his accompanying craft, but was spotted by the helicopter pilot as he had inflated a brightly coloured safety buoy.

The helicopter also flew over the border to help with the search for six children missing in Cushendall, Co. Antrim, but was stood down when the RAF took over.

ON 10 JUNE 2003, 24-hour helicopter search and rescue resumed at Waterford Airport, provided by CHC Helicopters on contract to the Irish Coast Guard. Several weeks later the Air Corps began providing a medium-lift rescue helicopter service for the first time from the north-west at Sligo Airport using a Sikorsky leased from CHC. The Air Corps crew had trained with CHC. After many wasted years of one-sided rivalry, there was now close co-operation between the two services. 'If there is one positive aspect to Tramore, one lesson that has been learned, it is this,' one of the pilots closely attached to both services observed. 'Ultimately this will benefit those in trouble on land and sea.'

The formal closing down of the base at Finner and its relocation to Sligo was low key, with no politicians or senior military personnel present — and no press. However, there was a sense of closure among the Air Corps crew on 14 July 2003 as the last Dauphin flight took off and flew across Sligo Bay, leaving behind the limestone bench dedicated to four former colleagues. Just four days before, at a High Court sitting in Waterford, the Department of Defence had admitted liability for the death of Captain Dave O'Flaherty and had agreed to a €1.1 million settlement for Maria O'Flaherty.[2]

It was the first time that liability had been admitted; in two previous cases, those taken by the families of Sergeant Mooney and Corporal Byrne, liability had been denied, while the case of co-pilot Captain Baker was still pending. The Byrne and Mooney families decided to write to the minister and seek an admission of liability, following the O'Flaherty judgment.

'A great sense of relief' was how Maria O'Flaherty described her feelings immediately afterwards. 'They did everything they could to attribute this to pilot error when that obviously wasn't the case.' The admission had come four years and eight days after the crash. At that point the four crew had not been given any official decoration by the Defence Forces. It was now time to give them their Distinguished Service Medals, Maria O'Flaherty said. And many in the Air Corps rank and file were right behind her.

20

GO MAIRIDÍS BEO[1]

In January 1988 a commercial aircraft passing over Dublin picked up a signal from a distress beacon. Air-traffic control was informed, and within minutes a Dauphin helicopter was flying east. However, the co-ordinates for the transmission from the satellite emergency position-indicating radio beacon (EPIRB) didn't quite match the Irish Sea. The vessel 'in distress' was on land, in the front garden of a private house in the southside suburb of Sallynoggin.

The bemused aircrew flew over the small blue and white angling vessel and returned to base. The next morning, the scramble phone at Baldonnel rang again for yet another EPIRB signal. It was the same craft and the owner was a mite embarrassed. 'I was wondering why the helicopter was hanging around so much,' he told the *Sunday World*. He had only recently towed the craft to his home and wasn't aware that an EPIRB had been fitted — and was transmitting, to boot.[2]

EPIRBs, developments in navigational equipment and improvements in safety equipment have done much to reduce risks at sea in the last decade, and many emergency taskings involve situations which could have been avoided if certain precautions had been taken. However, two men who were well prepared — and survived as a result — were sea anglers Bill Hepburn and John Gowan from Fife in Scotland.

The pair were spotted by the Irish Coast Guard Sikorsky helicopter from Dublin at about 3 p.m. on 1 September 2003 about 13 miles north-west of the northern tip of the Isle of Man. The alert had been raised the previous day when Liverpool Coastguard received a call from Fife police to say that relatives were concerned. The pair had not returned from an angling expedition on their 15 foot dory, the *Hell Raiser*.

Liverpool Coastguard sent rescue units to the Luce Bay and

Wigtown Bay areas of south-west Scotland and the Irish Coast Guard Sikorsky joined a Royal Navy helicopter in the sea search. Hepburn and Gowan were spotted clinging to their craft by the crew of the P & O passenger ferry, the *European Seafarer*, which serves the route between England and Northern Ireland.

'We got the shout that they had been seen, and we were only about ten minutes away,' the Sikorsky winchman Alan Gallagher says. The crew — pilot Captain Brian Brophy, co-pilot Liam Flynn, winch operator Derek Everitt, and Gallagher — had been flying for about two and a half hours at the time. When Everitt and Gallagher winched them on board, the two anglers were 'very happy men'. They had been some 25 hours in the water.

'They had gone out sea angling and they were anchored when the anchor rope fouled on the propellor,' Gallagher says. 'It pulled them round with the stern into the swell and they were swamped. The boat capsized but they managed to right it and they put on their lifejackets and neoprene dungarees. Then they got swamped again, the vessel went under — stern first — and would have sunk if it hadn't been for an airlock in the little half cabin.'

Fortunately the two men clung to the vessel. They had a far better chance of being seen, even if only part of the hull was visible above the waterline. They waved frantically as a number of vessels and one privately owned helicopter passed them by. One of them tied a rope around his body, ducked down and, in under the vessel, fished out some crisps, a banana and the vessel's emergency flares. Unfortunately the flares had become wet and wouldn't work.

'During the night the water kept spilling in over the tops of their dungarees, and so they had to makes holes in their boots to let the water out. When we came upon them they were sitting on a thin guard rail and complaining about sore rear ends,' Gallagher says. 'In fact one of them had himself tied on to it and had forgotten about that when I got to him. I had to cut him off when I realised he was snagged.'

The men were only mildly hypothermic after their ordeal because of their safety gear. They were flown to Nobles Hospital in Douglas on the Isle of Man. 'The fact that they had the survival gear is a lesson for us all,' Brophy said afterwards. 'In this day and age of search and rescue helicopters here and in Britain, once you have the right survival equipment you have a much better chance of being found, be it a day or so later.'

IRELAND, 2004. A STATE with the most globalised economy on earth, a right-wing coalition government keen to privatise State-owned enterprise, a widening gap between rich and poor, and a society where 'heroes' are entrepreneurs who run low-cost airlines, or highly paid international soccer players, or winners of reality television contests. Among senior politicians, the phrase 'do gooder' has almost become a term of abuse.

On 9 December 2003 lifeboats from Dun Laoghaire, Howth and Skerries made a rendezvous in Dublin Bay off Baily lighthouse. There was no major alert. The call-out was to mark 200 years of lifeboat rescue off the Irish coast. The RNLI estimated that at least 1,000 lives had been saved, and many hundreds more have been rescued from danger by voluntary lifeboat crews over two centuries since the first rescue boat was stationed at Sandycove, Co. Dublin, by the former Dublin port authority — the Ballast Board.

An Air Corps Dauphin helicopter flew overhead with photographers from the national news media. Significantly it was just over 40 years since the first search and rescue helicopters — the two Alouettes — had arrived into Baldonnel after an epic delivery flight from France. And it was almost half a century since the first recorded helicopter rescue in these waters by the British Navy.

Just over a week after the Dublin Bay commemoration with the lifeboats, and eight days before Christmas, the last remaining search and rescue base run by the Air Corps was informed that it was being 'stood down' permanently in early 2004. The news was conveyed to the pilots and crew in Sligo Airport by telephone, after a meeting between the Minister for Defence, Michael Smith, and senior military management. Due to a 'variety' of problems within the Air Corps, the service agreement with the Irish Coast Guard could not be met, the minister said.

Certainly, a form of industrial action by winching crews over pay and safety issues had restricted the Sikorsky's operation, and had scuppered hopes of delivering a 24-hour service from autumn 2003. When there was no resolution to the row, winching crews were assigned to non-flying duties back at Baldonnel. However, even the Irish Coast Guard appeared to be taken aback by the minister's decision; CHC Helicopters, providing sterling service on contract at the three bases at Dublin, Shannon and Waterford, had been asked to quote for a temporary service in Sligo which might buy time until the Air Corps had resolved its difficulties. It was always envisaged that a combination of State and private helicopter search and rescue was preferable to total dependence on one operator.

Fishing industry representatives, who were over in Brussels for annual fish quota talks, criticised the decision, given the close relationship built up over the years with the Air Corps in the north-west. Families of the four airmen who had died in Tramore, Co. Waterford, were also angry. The pilots and their families who had moved to Sligo, and participated in medium-lift training in Norway as part of an E11 million State investment in the north-west, were stunned. Several had signed on for further Air Corps service only weeks before.

It came as little surprise to the pioneers of search and rescue, the early Air Corps pilots and crew who had struggled with minimal resources in the early days of the Alouettes. Had the government been seriously committed to State involvement in air/sea rescue, efforts would have been made to address the Air Corps internal difficulties. As one pilot remarked, 'this is the culmination of years of official neglect'.

Up in Belfast, social worker and father of a newborn baby, Chris Rintoul, had another anniversary on his mind as he resolved to write several letters. These were not routine Christmas cards, but 'thank you' notes to four helicopter crew — all of whom have now left the Air Corps, though some are still flying.

For ten years Rintoul had lived with trauma, guilt and gratitude: trauma over his experience on Horn Head, Co. Donegal, in 1993; guilt for surviving the horrendous fall which claimed the life of his close friend, Peter King; and immense gratitude to the four aircrew who risked their lives to save him as he clung desperately to the scrub grass which was between him and a 250–300 foot sheer drop. His first child was born on 26 September 2003 — ten years and a day after that experience.

> It took me a long time to recover from it, and to cope with the fact that I made it and Peter didn't. For years I tried to blot it out, only to realise that I had to start acknowledging and remembering. I remember that I held on to the winchman's ribs so tightly that he was bruised afterwards — his injuries were worse than mine!'
>
> The aircrew literally put their lives out for me. I often think there could have been six dead there that day. I have them to thank for my entire life and that isn't something you can say to many people.

Not only were they courageous and highly professional, but they were also very sensitive — on the day and afterwards at the inquest into Peter's death. It is that sensitivity, that humanity, that I was struck by most . . .

NOTES

Chapter 1 (pp 1–5)
1. *The Donegal Democrat*, 7 March 1956.
2. Frank Shovlin, 'The day of the helicopters', *Dearcadh 2000–2001*.
3. ibid.
4. *The Donegal Democrat*, 7 March 1956.

Chapter 2 (pp 6–22)
1. John de Courcy Ireland, *Wreck and Rescue on the East Coast of Ireland*, Glendale Press 1983.
2. ibid.
3. *Lifeboats Ireland 1999: 175 Years of Duty*, Irish Marine Press 1999.
4. Lieutenant Colonel Michael O'Malley, 'In the beginning', *Irish Air Corps: Celebrating 30 Years of Helicopter Operations 1963–1993*.
5. Peter Whittle and Michael Borissow, *Angels Without Wings*, Angley Book Company 1966.
6. ibid, p. 51.

Chapter 3 (pp 23–44)
1. *The Irish Times*, 21 June 1967.
2. Mike Reynolds, *Tragedy at Tuskar Rock*, Gill & Macmillan 2003.
3. *Irish Independent*, 19 March 1970.
4. ibid.
5. *Evening Herald*, 15 July 1968.
6. ibid.
7. Frank Khan, *Irish Independent*, 31 March 1973.
8. ibid.
9. Rev. Thomas Rocks OFM, 'Rescue on Muckish', *Irish Air Corps: Celebrating 30 Years of Helicopter Operations, 1963–1993*, Captain David Swan ed.
10. 'That Others Might Live: The Air Corps Helicopter Rescue Service', presented and produced by Madeleine O'Rourke, RTÉ Radio One, 21 November 1996.
11. Rev. Thomas Rocks, 'Rescue on Muckish'.

Chapter 4 (pp 45–57)
1. Report by Pat Smyllie, *The Irish Times*, 9 January 1979.

2.   *The Irish Times*, 15 August 1979.
3.   *The Irish Times*, 25 June 1985.
4.   *The Irish Times*, 21 June 2000.

Chapter 5 (pp 58–66)

1.   Harvey O'Keeffe and Kevin Daunt, 'Helicopter training down the years', *Irish Air Corps: Celebrating 30 Years of Helicopter Operations 1963–1993*.
2.   Commandant Frank Russell, Officer Commanding Helicopter Squadron, 'Helicopter squadron history 1963–1984', *An Cosantóir*, March 1985.
3.   *Stornoway Gazette*, January 1988.

Chapter 6 (pp 67–81)

1.   West Coast Search and Rescue Action Committee Report, 1989, p. 13.
2.   The Irish Marine Emergency Service was later renamed the Irish Coast Guard.
3.   Air Corps Mission Report: Helicopter Search and Rescue Mission, 9 March 1990, off the Donegal coast, June 1991.
4.   ibid.
5.   ibid.
6.   *The Irish Times*, 6 April 1991.

Chapter 7 (pp 82–94)

1.   *The Irish Times*, 13 February 1992.
2.   Paul Beaver and Paul Berriff, *Rescue: The True-Life Drama of Royal Air Force Search and Rescue*, Patrick Stephens Ltd in association with Scottish Television, 1990, p. 55.
3.   Report of the investigation into the collision between the French FV *Agena* and the Irish FV *Orchidée* on 22 September 1992: published by the Minister for the Marine, September 1999.

Chapter 8 (pp 95–101)

1.   SAR Squadron No. 3 Support Wing: Mission report on search and rescue mission at Horn Head, County Donegal, 25 September 1993, by Captain Dave Sparrow.
2.   ibid., Sergeant Ben Heron.
3.   *The Irish Times*, 20 October 1993.
4.   As note 2.
5.   As note 1.

Chapter 10 (pp 112–21)

1. Report of the investigation into the loss of FV *Scarlet Buccaneer* at Howth, County Dublin, on 16 November 1995: published by the Minister for the Marine and Natural Resources, Dr Michael Woods, 10 March 1999.
2. Lieutenant Commander John Leech, Naval Service, 'In search of the *Carrickatine*', *An Cosantóir*, March 1996.
3. Lieutenant A. McIntyre, 'Search and rescue operations — offshore', *An Cosantóir*, March 1985, and see also pp 61–2.
4. As note 1.

Chapter 11 (pp 122–31)

1. *The Irish Times*, 27 October 1997.
2. Letters to the Editor, *The Irish Times*, 19 November 1997.

Chapter 12 (pp 132–7)

1. *The Irish Times*, 8 January 1998.
2. ibid.

Chapter 13 (pp 138–51)

1. Commenting in the official report on the reference to 'catching out' the crew in relation to response times, the director of the Irish Coast Guard (formerly IMES) said that the Coast Guard 'wishes to place it on record that, at all times and in respect of all taskings of declared facilities in the context of SAR incident management, the interest and safety of the crews are of paramount importance to the MRCC and inform all its decisions. No demands of an unreasonable nature nor any outside the terms of normal mutual understandings are made on declared facility crews. The Coast Guard very sincerely regrets the loss of the Dauphin crew in the accident at Tramore on 2 July following its completion of a successful SAR mission.'
2. Final Report Accident DH248/Rescue 111 Tramore at 00.40 hours local time 2 July 1999 by the Air Accident Investigation Unit, pp 27–28.

Chapter 14 (pp 152–66)

1. Miriam Lord, *Irish Independent,* 6 July 1999.
2. *The Irish Times*, 17 May 2003.
3. Final Report, Air Accident Investigation Unit, p 22.
4. Findings, Court of Inquiry Dauphin DH248, 30 August 2001, para. 30.

5. *The Irish Times*, 17 May 2003.
6. Findings, Court of Inquiry Dauphin DH248.
7. Letters to the Editor, *The Irish Times*, 6 August 2001.

Chapter 16 (pp 175–87)

1. Tony Bullimore, *Saved*, Little, Brown 1997.
2. Report into the sinking of the Irish fishing vessel *St Gervase* close to Mizen Head, County Cork, on 23 November 2000, Marine Casualty Investigation Board, September 2003.

Chapter 17 (pp 188–200)

1. Mission report on rescue of 13 people from fishing vessel *Zorrozaurre*, 30 November 2000: A Flight Number 22 Squadron, RAF, Chivenor, North Devon, 6 December 2000.
2. ibid.
3. ibid.
4. *The Irish Times*, 11 December 2000.
5. *The Irish Times*, 2 January 2001.

Chapter 18 (pp 201–10)

1. Ted Creedon in *The Kerryman*, 7 February 2002.

Chapter 19 (pp 211–17)

1. *The Irish Times*, 12 April 2003.
2. At the court hearing Maria O'Flaherty asked that the €25,400 sum for mental distress be shared with her daughter Davina, Captain O'Flaherty's mother Lily, and his siblings Dermot and Valerie.

Chapter 20 (pp 218–22)

1. 'Go mairidís beo' — that others may live — is the motto of Air Corps Search and Rescue.
2. The *Sunday World*, 31 January 1988.

GLOSSARY

CASEVAC – casualty evacuation

CO-PILOT – second pilot

DOUBLE LIFT – two-person hoist, usually the winchman and survivor

EPIRB – emergency position-indicating radio beacon which is activated on contact with water and transmits a latitude and longitude by satellite

FLIR – forward looking infra-red equipment

HI-LINE RESCUE – where the deck crew hold a line from the aircraft and the helicopter hovers to the side of the ship, often used in very rough sea conditions.

IMES – Irish Marine Emergency Service, which subsequently became the Irish Coast Guard

MAYDAY – international distress call which is only used when there is imminent danger to life

MEDEVAC – a medical evacuation

MRCC – Marine Rescue Co-ordination Centre attached to the Irish Coast Guard

PAN-PAN – international distress call when a boat is in difficulty, but the situation is not life-threatening or urgent enough to require assistance.

STROP – used to lift a survivor into the helicopter

TASKING – when a rescue unit is directed to an emergency situation by the relevant marine co-ordination centre

VHF – very high frequency (civilian radio). Channel 16 is the emergency communications channel.

WEAK LINK – a piece of light nylon line attached to the main hoist line with a breaking strain of 125 lbs. It serves as a safety device to protect the aircraft if the hi-line breaks.

WINCHMAN – airman lowered from the helicopter to the rescue scene

WINCH OPERATOR – crew member who controls the winching

INDEX

Aaron Stark Symes, 15
Aer Lingus, 25, 26–7, 139
Agena, 87, 89
Ahakista, 56–7
Ahern, Bertie, 147
Ahern, Dermot, 197
Air Accident Investigation Unit
    (AAIU), 145–7, 156–7, 163,
    165, 169
Air Corps
    ambulance service, 10–11, 23,
        24, 39, 58–9, 99–100, 123
    background, 19–22
    discontinued, 220–21
    ditching procedures, 88–9
    early role, 22, 23
    first missions, 6–10, 12–13
        fishery patrols, 28
    funding, 11–12, 14, 119, 162–3
    funding, inadequate, 109, 139
    government review (1996),
        119–20
    helicopters, 20–21, 60–61 *see also*
        Alouette; Dauphin
    hoax calls, 214
    industrial action, 220
    legal actions against personnel,
        214
    morale and staffing difficulties,
        140, 154–5, 162, 220
    mountain rescue, 23, 25–6,
        29–31, 39–44, 111
    privatisation of services, 139,
        220–21
    public awareness, 13
    safety record, 61 *see also* Tramore
        tragedy
    south-east region, 138–41, 155
    1,000th mission, 59–60
    training, 8–9, 11–12, 21, 60,

88–9, 195
    training deficiencies, 156, 157
    training, medical, 23–4, 99–100
    weather relief missions, 24, 53
    western region campaign, 67–70,
        71, 73, 138–9
air crashes, 25, 26–7, 53–7, 71, 139
    see also Tramore tragedy
Air India disaster, 53–7
Aisling, LE, 54–5, 103, 105, 116, 173
Alimar, 171, 172–3, 174
Allen, Paul, 160–61
Allen, Roderick, 36
Alouette helicopter, 9–10, 46, 60, 62
    commissioned, 6, 8, 11, 21, 26
    safety record, 61
    uses car petrol, 7, 9
Alvena, 51
ambulance missions, 10–11, 23, 24,
    39, 99–100
    babies, 58–9, 123
'Amhran Bréagach' (song), 185
Andrews, David, 107, 111
Angels Without Wings, 20
Angland, Timothy, 186
Anna Sophie, 62
An-Orient, 176–8
Aoife, LE, 61
Aran Island lifeboat, 7, 124, 178,
    183, 185
Aran Islands, 6, 23, 69, 71, 86, 105,
    123, 147–8
Árd Carna, 61–2, 116–17
Ardmore, 140, 167–9
Arosa, 175–6, 178–85
Arranmore Island, 74, 115
Arranmore lifeboat, 2, 3, 4, 61, 74,
    75–6, 78, 216
Association for Adventure Sports
    (AFAS), 23

Audley, Fr Padraic, 184
Aza, Alberto, 137

B & I, 82, 84–5
Baird, Gordon, 173
Baker, John, 153
Baker, Mick, 138, 140–41, 146–8,
    149, 152–3, 155–6, 217
Baker, Tony and Mary, 149, 152,
    155–6, 158, 161, 166
Baldonnel, 6, 8, 9, 19, 114, 119,
    158, 162
Ballycotton lifeboat, 17, 50, 143
Ballyglass lifeboat, 93–4, 125, 127,
    201, 202
Ballykelly, 117–18
Ballyvaughan, 79
Balmain, Captain, 2, 3, 5
Balthazar, 6
Baltimore, 214–15
Baltimore lifeboat, 50, 52, 186
Bambury, Jimmy, 207
Bantry Bay, 26, 45–7, 63, 108–9,
    134, 187, 192, 215
Barrett, Josie, 125, 127, 129
Barrett, Sean, 119–20
Batchelor, Andy, 189, 190–91
Bates, Willie, 27
Baulding, Mike, 79
Beechcraft helicopter, 61
Beginish Island, 71
Belderrig, 124–5
Belfast, 24, 197
Bell 212 helicopter, 46
Benbecula, 193, 194–5
Benchourn, 149–50
Bentley, John, 196
Bere Island, 108
Best, George, 26
Betelgeuse, 45–7
Big Cat, 70–71
Biger, Jacques, 186, 187
Blackburn, Colm, 198–9
Blacksod, 90, 149–50, 193–4,
    211–12
Bláth Bán, 126–7

Bleaden, John, 65–6
Bloody Foreland, 115
Blue Stack Mountains, 111
body scoop, 21
Bond Sikorsky, 139
Bonner, Shane, 167–9, 170–71,
    196–200
Bord Fáilte (boat), 31–6
Boulden, Alan 'Rocky', 54, 56, 88
'bousers', 9
Boyle, Denis, 113
Brabourne, Lady, 48
Bradley, Gerard, 128
Brady, Brendan, 24–5
Brady, Michael, 29–30
breeches buoy equipment, 71
Brennan, Lee, 203, 204–5
Broderick, Dr Marion, 67, 68, 99,
    105, 148
Brogan, Anne, 196, 201–3
Brophy, Brian, 219
Brown, Peter, 71
Brown, Terry, 54, 56
Brown, William, 113
Buckley, Sarah, 109
Bucks Fizz, 52
Bullimore, Tony, 177
Burke, Aidan, 187
Burns, Eamonn, 99, 129–31,
    179–80, 193, 194–5, 211
Byrne, Anna, 152, 166
Byrne, David (Blackie), 39–40
Byrne, Dinny, 64
Byrne, Gay, 59, 67–8
Byrne, Hugh, 89, 162
Byrne, Ken, 25, 31, 154
Byrne, Niall, 138, 140–41, 146–8,
    150, 152, 217
Byrne, Vincent, 150, 152, 162
Byrne, Willie, 42–3

Caher Island, 13
Cahirciveen, 37
Camargue, 52
Capall Bán, 178
Capitaine Pleven II, 78–81

Carcass, Vic, 66
Carey, Chris
    air-ambulance mission, 24
    on Alouettes, 62
    Emerance, 9, 10
    first sea rescue (Caher Island), 13
    Lugnaquilla air crash, 25–6
    rescue of diver, 26
    training, 11, 23
Carey, Paddy, 29, 36, 37, 48
Carlingford Lough, 48
Carlingford Mountain, 198
Carna, 176, 184
Carracelas, Euginio Diaz, 175, 183–4
Carrickatine, 112–13, 115–20, 138
Casburn, Bryan, 68
Casey, John, 29
Castletownbere, 27, 102, 109, 186,
        192, 215
Cathal Brugha, 24–5
Cavanagh, Charlie, 128
cave rescue, 124–7
Celestial Dawn, 205–10
Celtic Mist, 186
Challenger, HMS, 55
Chambers, Christopher, 5
Champion, Joe, 2
Chapman, Paul and Sean, 170–71
CHC Helicopters, 163, 216–17, 220
Chinook helicopter, 54, 56
Chivenor (Devon), 132
Churchill, Winston, 19
Ciara, LE, 201, 202
Civil Service No. 7, 15–16
Clancy, Sean, 166
Clark, Herwin, 110
Clarke, Joseph, 15
Clayton, Nick, 52
Cleggan Bay, 25, 201
Cleggan Coast Guard, 184, 185
Clew Bay, 19
Clifden, 7, 8
cliff rescues, 123–4, 129–31, 167–9
Cliffs of Moher, 123–4, 129–31
Clonan, Eugene, 184
Clyde, 192

Coast Life Saving Service, 71
Colleen, 12
Collins, Christy, 50
Collins, Maureen, 24
Comhairle na Mire Gaile, 71
Commissioners of Irish Lights, 185
Conlon, Fred, 36
Connell, John, 31
Coolaney, 90–91
Cooley Peninsula, 197–8
Corby, Jim, 74, 77, 92–3
Corcoran, Dave, 215–16
Corcoran, Joe, 91
Costello Bay Coast Guard, 184, 185
Cotter, Donal, 111
Cotter, Sean, 187
Cottrell, Simon, 129–30
Courtmacsherry lifeboat, 50
Courtney, David
    air-ambulance mission, 123
    An-Orient, 176–8
    Arosa, 179, 184–5
    career, 122–3
    cave rescue, 125, 128–9
    cliff rescue, 129–30
    Hansa, 193–5
    on Tramore tragedy, 164–5
Coy, Alan, 136, 137
Cranfield, Patrick (GOC Air Corps),
        138, 155, 157–8
Creedon, Ted, 206–9
Crilly, Niall, 27
CRM (crew resource management),
        164
Croke, Lewis, 19
Croke, Tom, 30
Crosby, Dane, 88
Crowley, Garda, 5
Crowley, Michael, 27
Crumlin, 100
Cumberland, HMS, 87–8
Cummins, Sergeant, 9
Currid, Carmel, 119, 121
Currid, Timmy, 114, 115, 118, 120,
        138
Curtin, Patrick, 71

Cushendall, 216
Cushnahan, John, 173

Daunt, Kevin, 60, 94, 150
Daunt Rock, 17
Dauphin helicopter, 69, 72, 93, 94,
    106
  commissioned, 62–3
  obsolescence, 155–6, 160–61, 162
  replacement tender, 162–3
  Tramore tragedy, 151, 154–8,
      160–61
Davies, Michael and Ros, 140
Davis, Mike, 159–60
de Almeida, Sebastian Vaz, 176, 185
de Costa Cravid, Albertino
    Herculano, 176, 185
de Courcy Ireland, Dr John, 14, 16,
    18, 69
de Paor, Rory, 25
de Yturriaga, Dr José A., 109
Deacon, William, 110
Defence, Dept of, 11–12, 14, 21,
    141, 156, 158–61
Deirdre, LE, 49–52, 107–8, 110, 116,
    173
Devereux, John, 18
Dingle, 205–10
Dingle Bay (boat), 207–8
Dingle lifeboat, 206, 207–8
Dodd, Steve, 173
Dodsworth, Flight Sergeant, 191
Doherty, Eamonn, 70
Doherty, Hubert, 61
Doherty, Hugh, 61
Doherty, Michael, 128
Doherty, Terry, 112, 113
Doherty report, 72–3, 139
Donnelly, John, 184, 185
Donnelly, Noel, 85–6, 124, 125,
    171–4, 177
Donovan, Dennis, 16
Doolin Coast Guard, 123–4, 129,
    131, 184
Doran, Robert, 86–7, 89
Dore, Cedric, 79

Dorrian, Paudge, 120
Douglas Hyde, 16–17
Downes, Kevin, 114
Downey, Ian, 115
Downpatrick Head, 31–6
Doyle, Ciaran, 126–7
Doyle, Eamonn, 68, 69
Dragonfly helicopters, 19–20
Dublin Bay, 82–3, 91–2, 220
Duggan, Billy, 17
Duggan, John, 17
Duggan, Richard, 17
Dun Laoghaire, 91–2, 204
Dun Laoghaire lifeboat, 15–16, 18,
    83, 220
Dunboy, 102–7, 110
Dundalk Bay, 197
Dungarvan, 140, 143
Dunluce, 196
Dunmore East, 138, 139, 140, 142
Dunmore East lifeboat, 50
Dunne, Alec, 13, 22, 28, 36
Dunne, Paul, 140
Durnin, Joseph, 25
Dursey, 102

Eagle Island, 19
Easkey, 24–5, 36–7
Eccleshall, Neil, 174–5
Eithne, LE, 62–3, 69, 71–2, 119,
    178, 183, 185
Emer, LE, 55
Emerance, 6–10
Endurance, 178
EPIRBs, 54, 70, 176, 186, 192, 218
EU coastguard, call for, 173
Eurocopter, 162–3
European Seafarer, 219
Everitt, Derek, 203–4, 219

Faherty, Mrs, 10–11
Fahey, Frank, 173
Fahy, Jim (Corporal), 8, 11
Fahy, Jim (RTÉ), 79
Fanad Head, 38, 115, 117, 216
Farrell, Brian, 206–7, 208

Fastnet, 49–52, 53, 187
Fenit lifeboat, 178
Finlan, Michael, 79
Finner, 72, 73, 114, 150, 158, 160, 165–6, 217
Finucane, Michael, 159
'fishbowl effect', 172, 188
fishery patrols, 28
fishing fleet, ageing state of, 119
Fitzgerald, Edmond, 140
FitzGerald, Dr Garret, 62
Fitzgerald, Mick, 11
Fitzgerald, Peter, 123
Flanagan, Aidan, 154–5, 166
Flanagan, Anthony, 82
Flannery, Jimmy, 207
Flynn, Liam, 219
Fóla, LE, 46
Forrestal, John, 94, 111, 115
Fort Descaix, 134
Forty Foot (Dun Laoghaire), 91–2, 204
Fox, Breda, 136
Fox, David, 134, 135, 136–7
Fox, Uffa, 20

Gallagher, Alan, 197, 219
Gallagher, John, 31
Gallagher, Pat 'the Cope', 66
Galtee, 2
Galway Bay, 78–81, 175–6, 178–81, 182–5
Garcia, Alfredo, 173, 174
Garcia, Ricardo Arias, 175–6, 179–81, 182, 183
Garda Sub-Aqua Unit, 125–7, 128
Gaunt, Denis, 54
Gawley, Donal, 84–5
Gazelle helicopters, 60
Gill, Austin (Kilkenny), 82–3
Gill, Austin (Ocean Tramp), 86
Gillougly, Gene and Elizabeth, 29–30
Gladonia, 71
Glendalough, 29
Glendalough (boat), 27
Glengad Head, 117

Glenmacnass waterfall, 31
'goldfish effect', 172, 188
Goldsberry, Dermot, 13
Goleen Coast and Cliff Rescue, 186
Goodbody, Robert, 171, 211
Gormley, Bernard, 112, 113
Gowan, John, 218–19
Gráinne, LE, 46
Gráinne Uaile diving club, 125, 127
Granton Osprey, 37–8
Granuaile, 115, 116
Greeley, Mark, 207
Green Lily, 110
Greencastle, 61, 112–13, 115, 119, 128
Greenhaven, 1–4, 5
Gregg, Dr Thomas, 24
Gregory Sound, 86
Gribble, Nick, 86, 103, 105, 110
Griffin, 'Skipper' Pat, 21–2
Groden, Tom and John, 86

Hackett, Emily, 167–70
Halpin, Nicholas, 108
Hamilton, Henry, 15
Hannah Pickard, 15
Hansa, 192–5
Harding, David, 84
Harrington, Kieran, 186, 187
Harris, Ian, 86
Hasselwerder, 82
Haughey, Charles, 64, 66, 86, 186
Haughey, Edward, 197
Hayes, Desmond, 84
Hayes, Paul, 90, 91, 92–3, 94
Heath, Ted, 52
Heffernan, Annamarie, 128
Heffernan, Michael, 124, 125, 126, 127–8, 129
helicopters, 20–21, 60–61 see also Alouette; Dauphin
Hell Raiser, 218–19
Helvick Head, 140
Helvick lifeboat, 142, 144, 150
Henry, Garth, 128
Hepburn, Bill, 218–19

Hernandez, Carlos, 211–12
Hernon, Coley, 105
Heron, Ben, 74, 75, 76–7, 95–101
Hestrul II, 51
Hickey, Dick, 17
HIFR (helicopter in-flight refuelling), 72
hi-line technique, 32–3
Histon, Paddy, 2
hoax calls, 214
Hook Head, 139
Hooper, Richard, 79
horizontal lift technique, 204
Horn Head, 95–9, 221
Horse Island, 124–7
Houlihan, Peadar, 71
Howth, 12–13, 15, 113–14, 120
Howth lifeboat, 12, 13, 84, 114–15, 220
Hudson, Andy, 66
Hughes, Kev, 137
Hughes, Nick, 128

Illustrious, HMS, 17
Ingham, Steve, 54
Inis Meáin, 23, 69
Inis Mór, 105
Inis Óirr, 123
Inishbiggle Island, 148
Inishbofin Island, 216
Inishkea Islands, 201–2
Inishowen Head, 61
Inishtrahull Island, 92–3
Inishturk Island, 10
Invention, 187
Irish Coast Guard, 157, 213, 220
Irish Helicopters, 8, 102, 107
Irish Independent, 29, 30, 152, 161
Irish Marine Emergency Service, 72–3, 84, 91, 102, 115, 119, 123
    Tramore tragedy, 139, 141
Irish Marine Search and Rescue Committee, 169
Irish Parachute Club, 13–14
Irish Skipper, The, 61–2

Irish Times, 9, 14, 45, 55–6, 84–5, 128
    Tramore tragedy, 153, 154, 155, 160, 163, 164
Isabella, 16
Isle of Man, 218–19
Iuda Naofa, 178
Ivers, Eddie, 61

Jackman, Brendan, 170
James, Steve, 193
Jenalisa, 138
John F. Kennedy, 18
John Robert, 16
Johnson, John, 64–6
Johnson, Steve, 88, 89
Jones, Gordon, 55
Joyce, Ciaran and Lavinia, 8
Joyce, Gerard, 185
Joyce, John (Corporal), 22
Joyce, John (rescued boatman), 31
Joyce, Pat, 8, 103
Joyce, Pete, 133, 134–5, 136, 137
Juan Carlos, King, 184
Juncal, Ramon Pardo, 175–6, 181–2, 184
Junger, Sebastian, 112

Kane, Gary, 186, 187
Kartli, 85
Kavanagh, John, 49, 50, 51, 52, 71–2
Kavanagh, John T.G., 25
Kavanagh, Martin, 126, 127
Kavanagh, Paddy, 68
Keane, Billy, 8
Kearney, John (Airman), 31
Kearney, John (*Ard Carna*), 61
Kearns, Squadron Leader, 20, 21
Kehoe, Patrick, 84
Kellett, Frank, 12–13
Kellett, Paul, 108, 109
Kelly, John, 112, 113
Kelly, John (J.P.), 31, 59, 62
    air-ambulance missions, 10, 23, 39

Kelly, John (J.P.), *continued*
    Emerance, 6, 8, 9
    Malcolm Crone, 37–9
Kelly, Mark, 206
Kelly, Michael, 125
Kelly, Patrick, 114
Kelly, Stephen, 112, 113
Kelly, Terry, 30
Kelly, Tom, 198–9
Kelly, Tomás, 178
Kenmare, 27–8
Kennedy, Tom, 207
Kilkee Coast Guard, 184
Kilkeel, 197
Kilkenny, MV, 82–5
Killala Bay, 93
Killala Coast Guard, 125
Killeen, Mick, 1–2, 4
Killybegs, 64, 119, 173, 181, 213
Killybegs Fishermen's Organisation, 66, 67, 68, 166
Kilmore Quay lifeboat, 16, 18–19, 27
King, Peter and Raymond, 96, 98, 99, 221, 222
Kirby, Carmel, 102–6, 107, 110
Kirwan, Dara, 185
Kirwan, Jim, 59, 60
Kirwan, Liam, 84, 103, 120, 127
Kish lightship, 12, 13
Knatchbull, Nicholas and Timothy, 48
Kowloon Bridge, 63–4, 72, 73
Kruse, David, 114, 115

Lacken, 25, 128
Lady Christine, 215–16
Lady Murphy, 18–19
Lambay Island, 15
Laragh, 31
Lavelle, Tony and Anthony, 201–3
Le Borgne, Jean Marc, 79
Le Drian, Jean Yves, 80–81
Leauté, Xavier, 177–8
Leech, John, 116
Leinebjorn, 148

Leinster, MV, 83
Leonard, Peter, 104–5, 110, 123, 124, 177
    Celestial Dawn, 205–6, 208–10
Lewis, Hayden, 173
lifeboat service, 14–19, 220
Lioran, 185
Lister, Mark, 193
Lloyd's List, 213
Locative, 74–8
Lockey, Al, 86, 123
Loo rock, 78–81
Loop Head, 103, 176
Lord, Miriam, 152
LOST (Loved Ones of Sea Tragedies), 119, 121
Loughnane, Donal, 43, 48
Lugnaquilla mountain, 25–6
Lynch, Bernard, 27
Lynch, Dick, 59
Lynch, Joseph, 31
Lyne, Nealie, 55
Lynott, Patrick, 24–5
Lyons, Eileen, Tom and Tomasina, 58–9

McAdam, Neil, 91–2
McAleese, Mary, 147, 151, 152
McAvock, Padraig, 125
McBride, J.J., 5
McCarry, Sean, 128
McCartney, John, 91, 95, 96–9
McCombie, Thomas, 15
McDermott, John, 53–4, 102–7, 110
McDonagh, James, 5
McElroy, Brendan, 37, 38–9
MacFarlane, Ian, 62
McGinley, Joan, 67–8, 73, 107, 109, 139
McGrath, John, 54, 56
McGrath, John Tristan, 203–5
McGrath, Martin, 202
McGurk, Paddy, 59–60
McHale, Sean, 125, 126, 127
McHale, Tom, 24–5
McKeon, Sean, 1–4

McKinney, Conal, 112, 113
McKinney, Jeremy, 112–13
McLoughlin, Dr Ciaran, 184
McLoughlin, Joseph, 12
McMahon, Barney
    air-ambulance missions, 10–11,
        23, 24
    career, 28–9, 31, 46
    first mission (*Emerance*), 6–9
    first sea rescue (Howth), 12, 13
    Lugnaquilla air crash, 26
    mountain rescue, 29–31
    RAF co-operation, 26
    Tuskar Rock, 27
McNamara, Tony, 93–4, 125–6, 127,
    128–9
McParlin, Philip, 119
McVitty, Desmond, 5
Madden, Fr Brendan, 151
Mahady, Christy, 78, 111
Maher, Gerry, 54
Mahon, Muiris 'Mossy', 54, 56
Malcolm Crone, 37–9
Malin Coast Guard, 95, 173, 193,
    215
Malin Head, 112, 115, 117
Mangouras, Apostolos, 213
Manning, John, 125, 128, 171, 180,
    206
Manrique, Andres, 211–12
Mapescal, 134
Mar de Los Sargazos, 187
Mara Sul, 103, 107
Marine, Dept of the, 64, 70, 72–3,
    83, 156, 159, 161, 162
Marine Accident Investigation Branch
    (MAIB), 182–3
Marine Rescue Co-ordination Centre
    (MRCC), 36, 46, 53, 69, 71,
    72–3, 83, 91
    Tramore tragedy, 142–3, 156
Marry, Keith, 82–3
Mary Joseph, 139
Mary Stanford, 17
Maughan, Dr Eoin, 161
Maxwell, Paul, 48

Maybin, Michael, 12–13
Meehan, Benny, 170
Melampus, HMS, 15
Melchior, 6
Mellotte, Peter, 30
Melmore Head, 37
Meston, John, 137
Michael, Richard, 189
Milford Eagle, 171–4
Mills, Martin, 31
Mine Head, 62
Mizen Head, 124, 188–9, 192
Moloney, Thomas, 156
Monks, Fr Frank, 152
Monks, Kieran, 72
Mooney, Monica, 149, 153, 166
Mooney, Paddy, 138, 140–41, 146–8,
    149–50, 153, 217
Moonstone, 51
Morning Cloud, 52
Moss, Arthur, 52
mountain rescue, 23, 25–6, 29–31,
    39–44, 111
Mountbatten, Lord Louis, 48
Mountcharles, 170
Muckish Mountain, 40–44
Mulhall, David, 126, 127
Mulhern, Kevin, 198
Mullaghmore, 47–8
Mullins, Alan 'Arfur', 189, 190–91
Mullins, Patrick, 185
Mulroy Bay Coast and Cliff Rescue
    Service, 95
Murphy, Ciaran, 167–9, 170
Murphy, Martin, 114–15
Murphy, Neville, 211, 212, 214–15
Murphy, Peter, 68
Murphy, Sean (Commandant), 78–9,
    100, 114, 115, 148, 149–50
Murphy, Seanie (lifeboatman), 55
Murphy, Tish and Niall, 202
Murphy, Tony, Carmel and Emma,
    124–7
Murray, Des, 196
Murray, Dick, 37, 40, 62
Murrin, Joey, 66, 67, 68

Murvagh, 170
Myers, Kevin, 128

Naomh Eanna, 123
Narin, 1, 4–5
Neptune, 67
Nequest, Derek, 171–3, 206, 209–10, 214
Neves, Fernando, 178
New Quay, 28
Ní Fhlatharta, Mairéad, 123
Norbrook Laboratories, 197
Nuestra Senora de Gardotza, 107–10, 192
Nueva Confurco, 214–15
Nuska, 195–6

Oakes, Sean, 11
Oakley class, 18–19
O'Carroll, Declan, 166
Ó Cathain, Máirtín, 185
Ocean Tramp, 85–6
Ó Cearbhalláin, Daithi, 74, 75–78, 79–80, 91–2, 94
O'Connell, Bill, 11, 13
O'Connell, Sean, 126, 127
O'Connell, Fr Tony, 152–3
O'Connor, Fergus, 31, 91
    air-ambulance missions, 10, 23
    Betelgeuse, 45, 46
    Emerance, 6, 8–9
    on public awareness, 13
    Redbank, 28
    Sea Flower, 27–8
    Tipperary floods, 22
    training, 11, 23
    west coast action committee, 68, 69
O'Connor, John, 165
O'Connor, Dr Joseph, 161
O'Connor, Kevin, 29
O'Connor, Michael, 71
O'Connor, Tom, 91–2, 115, 117–18, 166, 204
Ó Donnchadha, Fr Gearóid, 178
O'Donnell, Donal, 64–5

O'Donnell, Hugh, 31, 36–7
O'Donnell, John, 76, 78
O'Donnell, Martin, 126–7
O'Donnell, Patrick, 126–7
O'Donnell, Tony, 184
O'Donoghue, Gerry, 72
O'Donovan, Derry, 109
O'Flaherty, Dave, 138, 140–47, 148–9, 151, 166, 217
O'Flaherty, Maria, 138, 141, 151, 153, 160, 162, 166, 217
Oglesby, John and Martin, 67
O'Hehir, Michael, 53
O'Higgins, Kieran, 213
oil, effect on wave action, 112
O'Keeffe, Brian, 161
O'Keeffe, Harvey, 60, 78, 79
O'Keeffe, Hugh, 25
Okori, Alberto, 195–6
O'Leary, Donal, 13
O'Mahony, Tony, 211–12
O'Maille, M.J., 4
O'Malley, David, 36
Onassis, Aristotle, 17
O'Neill, Jim, 148, 150, 197–8, 215
O'Neill, Willie, 13
Orchidée, 86–9
O'Regan, Michael, 124
Orla, LE, 88, 192
Ormsby, Paul, 173
O'Rourke, John, 148
O'Rourke, Madeleine, 43
O'Shaughnessy, Jim, 31
O'Shea, Paddy, 39–40, 42–3
O'Sullivan, Gene, 192
O'Sullivan, Gerry, 107
O'Sullivan, Richard (Dick), 43, 91, 106
Our Lady of Lourdes Hospital, 23–4
Overijssel, 50

Palme, 15
Paraclete, 64–6
Pearce, David, 36
Pembroke, 192
Perfect Storm, The, 112

Pierce, Kenneth, 86–7
Pile, Chris, 193
Poolbeg lifeboat, 15
Portnoo, 5
Portrush lifeboat, 61, 196, 198
Portstewart, 198–200
Potter, Alan, 132–7, 147, 152
Power, Jimmy, 86–7, 89
Power, Tracey, 89
Powerscourt, 30
Prestige, 213
Price Waterhouse review, 154, 157
Princess Eva, 211–13, 215
Puma helicopter, 52–3

Quinn, Michael (air traffic control), 53
Quinn, Michael (Leading Seaman), 107–8, 109–10, 192

Radiant Way, 94
RAF *see* Royal Air Force
RAF Escaping Society, 191
Rathlin O'Beirne Island, 2, 74, 115
Rawdon, Tom, 67
Redbank, 28
Regardless, 50, 52
Reilly, Gerard, 127
Renmore barracks, 7, 9
Representative Association of Commission Officers, 160–61
Richard, Brother, 40–41
Ring, John, 30, 36
Rinnishark, 203
Rintoul, Paul, Chris and Alan, 96–9, 221–2
Rivadulla, Luis Miguel Vidal, 184
RNLI (Royal National Lifeboat Institution), 11, 14–19, 138
road accidents, 24, 100, 170
Roancarrig Mór, 108
Roaninish Island, 1–4, 5
Robertson, George, 137
Robinson, Grant, 36–7
Robinson, Jim, 55, 56
Rocks, Fr Thomas, 40–42, 43–4

Rodgers, Brendan, 174
Rodgers, Gabriel, 29
Rogers, Edmund, 31
Rohan, Ken, 50
Roisin, LE, 186
Roleston, Noreen, 125
Roper, Joe, 166
Rossaveal, 182, 184
Rossiter, P.J., 140
Rosslare, 11, 139
Rosslare ferry, 159–60, 213–14
Rosslare lifeboat, 16–17, 27, 88, 214
Royal Air Force (RAF), 19, 45, 62, 79, 147, 188
    Air India disaster, 54, 55
    co-operation, 26, 118
    Doherty report, 73
    helicopters and, 20–21
    RAF Escaping Society, 191
    Sonia Nancy, 132–7
    training for Air Corps, 8, 11
Royal National Institution for the Preservation of Life from Shipwreck (RNIPLS), 14
Royal National Lifeboat Institution (RNLI), 11, 14–19, 138
Royal Navy, 3–4, 19–20, 45–6
Royal St George yacht club, 11
Rudden, Martin, 61
Ryan, Eugene, 183, 184
Ryan, Michael, 95, 98

St Cathal, 25
St Columba, 108
St Gervase, 186–7
St Kevin's Bed, 29, 39–40
Saint Killian, 50
St Theresa, 201
Salmon, Stephen and Christopher, 30
Saltee Islands, 139
Sanchez, Pedro, 211–12
Saunders, Ian, 133, 134, 135–6, 137, 188–91
Scanlon, Donal, 106
Scarlet Buccaneer, 114–15, 119, 121

Sceptre, 187
Sea Eagle, HMS, 26
Sea Flower, 27–8
Sea King helicopter, 54, 56, 69, 88
Sealink, 108
Sean Foy, 36
Seán Pól, 64
Second World War, 19–20, 191
Setanta, LE, 46
Shadow V, 48
Shannon, 30, 73, 78, 139
Shannon, Mattie, 124, 131
Shaw, Mike, 177
Sheehan, John Michael, 27
Sheehan, Noel, 27
Sheehy, Patrick, 136
Sheeran, Peter, 6, 10, 11
Sheridan, Liam, 11
Sherry, Owen, 42–3
Shetlands, 110
Shipwrecked Fishermen and
    Mariners' Benevolent Society, 80
Shovlin, Nora and Frank, 1, 4, 5
Sikorsky, 162–3
Sikorsky helicopter, 106, 107, 139,
    195, 215
Silver Apple, 51
Silver King, 216
Sinéad, 126
Sinnott, Fintan, 18, 19
Skeena, 29
Skerd rocks, 175–6, 178–81, 182–3,
    185
Skerries lifeboat, 15, 220
Slater, Ian, 79
Slebech, 187
Sligo Airport, 216–17, 220–21
Sligo Bay, 93, 215–16
Slyne Head, 103
Smith, John, 198
Smith, Michael, 110, 220
    Tramore tragedy, 147, 150–51,
        153, 154, 155, 158–9, 163
Smith, Peter, 11
Smyth, Ciaran, 198–9, 215–16
Soares, Orlando, 176, 185

Solly, Rose and G.C., 111
Sonia Nancy, 132–7, 147
Spanish fishing boats, 102, 171, 175,
    178, 181, 187, 188, 192, 195,
    214
    criticised, 107, 109, 137
Sparrow, Dave, 58–60, 89–91, 95–9,
    100–101
Sperin, Jim, 55–6
Star (newspaper), 158
Stena Europe, 213–14
Stephens, Mark, 79
Stirling, Robert, 128
Stornaway, 193, 196, 197
Swan, Dave, 111
Sweeney, Vincent, 90, 193

Tearaght, 15
Tergiste, 15
Thompson, Aidan, 167–9, 196
Tipperary, flooding, 22
Tobin, Pat, 140
Todd, Jeff, 65–6
Toe Head, 63
Tompkins, Gerry, 125
Tory Island, 92, 94, 116
Tpocachi Star, 189
training, 8–9, 11–12, 21, 60, 88–9,
    195
    deficiencies, 156, 157
    ditching procedures, 88–9
    medical, 23–4, 99–100
Tralee, 58
Tramore Bay, 71, 203–5
Tramore tragedy, 138–51
Tramore tragedy aftermath, 147,
    150–66
    inquest, 161
    investigation, 147, 154, 155,
        156–9, 160–61
    liability denial, 159
    liability settlement, 217
    military court of inquiry, 163–4
    replacement aircraft row, 153,
        154–6, 157–8, 162–3
    safety audit, 162

safety recommendations, 157, 158–9, 160–61
Treacy, Mick, 47–8
Treacy, Raymond, 47
Trinity House, 20
Tuskar Rock, 26–7, 139, 213
Typhoon, 72
Tyrrell, Jack, 18

Union Hall, 187
up-and-over recovery, 34–5

Valentia, 37, 71, 173, 175, 178
Valentia lifeboat, 27, 55, 71, 206, 208
Verling, Donagh, 201, 215
Versl, Dietrich, 124
von Below, Will Ernest, 124–5, 127

Wall, Jack, 159
Walsh, Aidan, 71
Walsh, James, 17
Walsh, Owen, 71
Walsh, Richard, 17
Walsh, Thomas (lifeboatman), 18
Walsh, Thomas (rescued sailor), 31
Ward, Fr Alan, 166
Ward, Davitt, 203–5
Warren, George, 5
Waterford Airport, 138, 140–45, 155, 156–7, 158, 159, 164
    CHC helicopters, 163, 216
    closure threat, 153
Watson, Dave, 133, 137
wave action, 112
Waveney class, 17–18
weather relief missions, 24, 53
Weaver, John, 28

Webber, W.H.J., 191
West Coast Search and Rescue Action Committee, 67–70, 71, 73, 138–9
Whelan, Andy, 117–18
Whiddy Island, 26, 45–7, 63
White, Brendan, 31
White, John, 115
Whittle, Peter and Borissow, Michael, 20, 21
Whyte, Jurgen, 74, 75–6, 77, 78
Wickham, John, 17
Wicklow mountains, 29–31, 39–40
Williams, Peter, 79
Williams, William W., 17
Wilson, John, 80
winch operation
    hazards, 38–9, 65–6, 75–8, 95–8, 104–5, 190, 199
    hi-line technique, 32–3
    strop design, 11, 20
    technological improvements, 110
    training, 11–12
    up-and-over recovery, 34–5
Winks, Gerard, 51
Wizard, 2, 3, 4
Woods, Dr Michael, 89, 120–21, 127, 147
    development of service, 138, 139, 140
World Concord, 16–17
Wright, Jeremy, 198

Yarrawonga, 71–2, 73
Youghal lifeboat, 167

Zorrozaurre, 188–92